Understanding inherited disorders

Understanding
inherited disorders

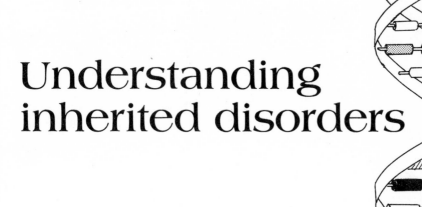

LUCILLE F. WHALEY
R.N., M.S.
Associate Professor, San Jose State University,
San Jose, California

with 121 illustrations

THE C. V. MOSBY COMPANY
Saint Louis 1974

Copyright © 1974 by The C. V. Mosby Company

All rights reserved. No part of this book may be reproduced in any manner without written permission of the publisher.

Printed in the United States of America

Distributed in Great Britain by Henry Kimpton, London

Library of Congress Cataloging in Publication Data

Whaley, Lucille F   1923-
   Understanding inherited disorders.

   Includes bibliographical references.
  ✓1. Medical genetics. ✓I. Title.
[DNLM: 1. Genetics, Human. 2. Hereditary diseases.
QZ50 W552u 1974]
RB155.W46   616'.042   74-1010
ISBN 0-8016-5418-1

CB/CB/B   9 8 7 6 5 4 3 2 1

To my family

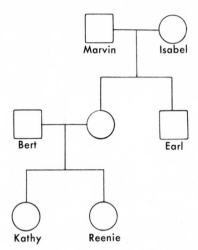

Preface

It is not difficult to justify a resource book on the basic concepts of genetics as applied to the health sciences. The importance and broadening scope of this branch of science is evidenced by the professional books and articles on the topic that are appearing in increasing numbers. The relevance to health-related disciplines is reflected in the expanded chapters on human heredity appearing in new and revised textbooks as well as in articles in a wide variety of current journals.

There is no doubt that inherited diseases constitute a significant proportion of world health problems. Their importance and relevance are recognized by allied health professionals, such as nurses, social workers, physical therapists, occupational therapists, health educators, teachers of handicapped children, and many others who continually encounter problems related to inherited defects and disease. This book was written in response to requests from many of these interested and concerned persons who seek to increase their knowledge of hereditary disorders but who find available volumes on medical genetics too complex and lengthy for practical use. It is hoped that this book meets this need by introducing the reader to the basic concepts of heredity as they apply to health situations. For the health practitioner it provides a ready resource; for the student, a supplement to standard texts. Hopefully, it may stimulate many to share an enthusiasm for this fascinating and rapidly changing field of science and a motivation to expand their knowledge and understanding with further study and investigation.

The book is concerned exclusively with human variations. The format is designed to assist the reader to better understand the fundamental mechanisms of heredity by first presenting the principles, then the exceptions and application. The first two chapters deal with the basic principles of heredity that provide a background for the remaining seven. The next five chapters are primarily concerned with various types of disorders as they apply to individuals and their relevance to health and well-being. Although the emphasis of the book is on the genetic principles, also included are symptoms, diagnosis, and treatment as they relate to the hereditary element, and in each category a representative example has been selected for elaboration to illustrate treatment and supportive care. Chapter 8 introduces the reader to the concepts of inheritance on a broader scope as they apply to populations; the final chapter is devoted to the application of genetic principles in treatment of genetic disorders and in genetic counseling.

A glossary of the more common genetic terms used throughout the text has been provided to facilitate reading of the content. Selected references are included at the end of each chapter, and the interested reader is directed to the excellent medical and biology texts available on the subject, some of which are included in the general references list.

Although the responsibility for any errors or omissions in the final manuscript is mine, I wish to express my thanks to those who have helped or inspired me in the writing of the book. My colleagues Gloria Vanisko and Margaret Jensen were kind enough to read the manuscript and offer suggestions and encouragement. I owe a debt of gratitude to Dr. Louis Levine of the City College of New York for the time he spent in carefully reviewing the manuscript. Very warm and special thanks go to my daughter Maureen, who transformed a jumble of cut and pasted pages into a typed manuscript; to my daughter Kathleen, whose collaboration was essential in the tedious preparation of the illustrations; and to my devoted husband, Bert, whose support and forebearance made the project possible.

LUCILLE F. WHALEY

Contents

4 CHROMOSOMAL ABERRATIONS, 73

5 GENETIC AND ENVIRONMENTAL INFLUENCES ON DIFFERENTIATION AND DEVELOPMENT, 90

6 GENES AND IMMUNITY, 114

7 INHERITANCE OF COMMON DISEASES AND DISORDERS, 136

Introduction

Man has always observed that "like begets like." Cats always have kittens, never puppies. Offspring will resemble the parents yet will differ from them and from one another. No two organisms are ever exactly alike. The similarities and differences between individuals result from the interaction of two distinct agents: genes and environment. The human organism begins his existence with a physical, biochemical, and mental potential determined by the genes he receives from each of his parents. Equally influential in shaping the individual throughout his lifetime is the environment, which is neither constant nor dependable. Rarely can one be fully separated from the other. *Genetics,* the science of inheritance and variation, attempts to explain these similarities and differences among living organisms and differentiates between what is inherited and what is not. Acquired differences cannot be transmitted from parent to offspring; inherited variations in the parental generation are likely to appear in offspring or in subsequent generations.

Human genetics is concerned with those characteristics that distinguish human beings from nonhuman beings; with those traits that characterize certain groups of individuals, certain families, and certain individuals; and with the way in which genes are expressed in the development of the individual during a lifetime. It is also concerned with the way the environment influences the individual throughout his lifetime.

Pope has written that "the proper study of mankind is man." Man's engrossing interest in himself has facilitated research into his similarities and differences, yet the difficulty of studying man as an experimental animal is apparent. A mouse can complete a generation in two months, a fruit fly in two weeks, and some microorganisms in a matter of minutes; man normally requires 20 years to complete a generation. A mouse produces scores of offspring during its lifetime and the fruit fly hundreds; the number of offspring man produces in a lifetime is small. Therefore, man's knowledge about himself has been gleaned primarily through observation of nature's "mistakes."

Long before genetics became a science, simple patterns of inheritance were known but not understood. Obvious traits such as extra digits and hemophilia could be traced from parent to offspring, and some early investigators even described the essential features of inheritance in these disorders. Two early scientists who had impact on the development of genetics are Charles Darwin, who upset the scientific world with his publication *Origin of the Species,* and Francis Galton, who suggested in 1876 the twin method for separating the effects of heredity and environment and began the study of multifactorial inheritance. But it was the Austrian monk Gregor Mendel who discovered the basic mechanism of heredity. The importance of his experiments was not recognized at the time and lay unnoticed in the scientific literature for almost 40 years until 1900, when three scientists working independently rediscovered Mendel's pioneer work on heredity. Genetics assumed its place as a field of science and has grown by leaps and bounds.

Much of the credit for rapid advances in the early part of this century is due to the work of Thomas Hunt Morgan with the tiny redeyed fruit fly *Drosophila,* whose rapid rate of reproduction and minimal requirements make it an ideal organism for experimentation. This tiny organism probably contributed more to the study of inheritance than all other sources during this period when thousands of experiments and countless discoveries were reported.

The English pediatrician Sir Archibald Garrod initiated the study of biochemical genetics when he interpreted the relationship between genes and some chemical reactions and published these observations in his well-known *Inborn Errors of Metabolism.* Unfortunately, as with Mendel's work, his findings were ignored for years. He now ranks with Galton as a founder of the subdivision of genetics known as medical genetics.

The rapid strides in the understanding of relationships between genetics and disease have been made through the efforts of scientists and their discoveries such as the ABO blood group system, the positive relationship between genes and infectious disease (malaria and sickle hemoglobin), the recognition of gene linkage, positive identification of specific enzyme deficiencies, and the role of genes in determining the structure of proteins.

The elucidation of the chemistry and structure of DNA and RNA and cracking of the genetic code are milestones in genetic research. Cytogenetics has provided information regarding the activity and structure of chromosomes, including the true number of chromosomes in man (46). The study of chromosome abnormalities has provided an explanation for many disorders, the most notable of which is Down's syndrome, or mongolism.

GENETICS AND DISEASE

All human characteristics have a genetic component including those characteristics interpreted as disease. Since most deviations from the normal are deleterious, they are brought to the attention of health workers. Some diseases are produced through the action of genes inherited from parents; others are acquired through the action of the environment on the genetic makeup of an individual. The genetic component in some disorders is obvious; in others it is subtle or scarcely discernible. Some disorders are apparent at birth; others may appear at various stages of the life span. It is difficult to estimate the numbers of persons whose lives will be limited or altered by genetically related defects or disorders.

The significance of heredity as an etiologic factor in disease is not a new concept although its importance to medicine and health has not always been obvious. Until relatively recent times the most important cause of death was infectious disease, particularly in the young. Nutritional disorders also took their toll, decreasing the host's resistance to infectious agents. The life span was short, and inherited disorders of late onset were transmitted to future generations before becoming apparent in previous ones. Persons with many inherited disorders for which there was no therapy died at an early age. As the effect of infectious disease and nutritional disorders has decreased there has been a corresponding increase in the importance placed on genetic defects as a cause of disease. Because of identification and treatment of hereditary disease, persons affected with genetically determined disorders are, with appropriate therapeutic intervention, now able to live a relatively normal life span.

Every year numerous biochemical and structural abnormalities are identified as the cause of genetic defects and disease. Health personnel are increasingly involved with the detection, treatment, and education of affected persons and their families. Widespread and comprehensive public health programs are attempting to provide services directed toward meeting these needs. As the number of persons requiring these services increases, the need for prepared professionals will also increase.

It is essential that professionals involved in the health care of persons with genetically determined or influenced disorders have an understanding of the nature of hereditary defects, the therapy, and the way in which these disorders will affect others. Clients are seeking information related to specific disorders and the problems they and their families can anticipate. Public knowledge is providing these clients with current information, and they will expect the health worker to provide answers to their questions.

1

The physical basis of inheritance

The information that determines the physical, chemical, and mental characteristics of an organism and provides for the orderly and consistent transmission of hereditary material from generation to generation is contained within the nuclei of the male and female reproductive cells. Although the outcome is continually influenced by environment, the potential of any individual is firmly established by his hereditary composition. "Genetics is to biology what the atomic theory is to the physical sciences."[3] The concept that forms the basis for the science of genetics is the gene theory of inheritance.

THE GENE THEORY

The method by which hereditary material is transferred and maintains its identity from one generation to the next is known as the *gene concept* or *gene theory*. The physical and chemical individuality of an organism is determined by the action of tiny units or particles called genes. *Genes,* occurring in pairs, are derived in equal parts from the father and mother during the process of reproduction. Each cell contains thousands of genes and each controls or regulates a specific cellular function. Alone or in conjunction with other genes, they are responsible for all the traits or characteristics seen in the organism; collectively they are responsible for continuity of the species and for orderly development and dynamic equilibrium of the individual.

The genes themselves are too small to be seen even through a microscope, but the effect they produce can be observed and analyzed. Until recently, most of the information accumulated regarding gene action had been gained through observation of physical alterations produced by alterations in the genes. It is now known that these physical manifestations of gene action are dependent upon chemical reactions at the cellular level.

The term used to describe the gene constitution of the individual is the *geno-*

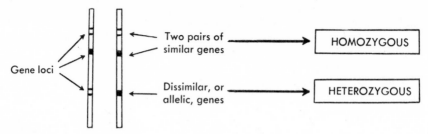

Fig. 1-1. One pair of homologous chromosomes with similar and dissimilar genes at three gene loci.

type. The appearance or observable characteristics of the individual that result from the interaction of the environment on the genotype is termed the *phenotype* (Gr. *phainein,* to show, appear; *typos,* form). Factors that alter the genotype will produce an effect on the phenotype even though the relationship between the two seems remote.

The genes are arranged in linear order along highly organized structures within the cell nucleus, the chromosomes (an arrangement that has been likened to that of beads on a string). All somatic (body) cells contain two sets of chromosomes collectively bearing two matching sets of genes, one set being contributed by each parent. Since they carry similar genes, each member of a pair of matching chromosomes is termed *homologous* (Gr. *homos,* the same; *legein,* to speak). Each gene has a definite position, or *locus* (pl. *loci*), on a specific chromosome and may take one of several different forms; for example, a gene for eye color may form eyes of brown, blue, or green. These alternate forms of a gene are termed *allelomorphs,* or *alleles* (Gr. *allelon,* of one another; *morph,* form). When corresponding gene pairs produce the same effect they are said to be *homozygous* (Gr. *homo,* the same; *zygotos,* yoked). When genes of an allelic pair produce different effects they are said to be *heterozygous* (Gr. *hetero,* different) (Fig. 1-1).

During germ cell *(gamete)* formation the chromosome number is reduced to a single set so that, when combined with the single set of chromosomes of the partner gamete, the resulting organism receives the full complement of genetic material. It is the interaction between the genes on two sets of chromosomes that produces the individuality of the organism. Although the genes themselves remain intact, the various chromosomes separate from one another during gamete formation to recombine with their homologous chromosomes during reproduction. Chance alone determines which member of a chromosome pair appears in the gamete and which gametes combine to form a new individual. These processes will be elaborated upon in the following section on chromosome behavior and in the section on principles of inheritance (p. 25).

THE CELL

The human body is composed of millions of cells that vary in size and shape according to their location and function. A brief outline of their major components will serve to place in perspective those structures that are most directly

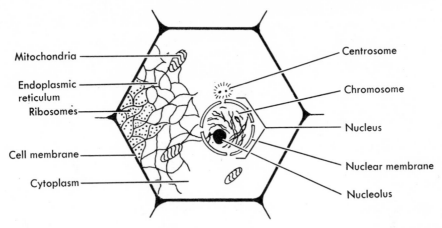

Fig. 1-2. A somatic cell.

concerned with inheritance. Enclosed within the cell membrane are the cytoplasm and the nucleus (Fig. 1-2). The nucleus, located in the center, is separated from the cytoplasm by a double membrane that encloses its components: the nucleolus and the chromosomes, so named because they stain a darker color. Scattered throughout the cytoplasm are a variety of *organelles,* some of which are the *mitochondria,* essential for energy production; a *centrosome* containing two centrioles, which play an important roll in cell division; the *ribosomes,* which are the site of protein synthesis; and the *endoplasmic reticulum,* a network of tubules or canals to which the ribosomes are attached. These organelles are extremely small and can be examined only with the electron microscope.

The cell components primarily concerned with heredity are the nuclear structures, the chromosomes. Chromosomes and their activity have been studied in both plants and animals for over a hundred years. With new equipment and techniques, human chromosome activity has become a highly specialized area of study. The branch of science that deals with the study of chromosomes in relation to associated genetic changes is called *cytogenetics.*

The chromosomes

There is considerable variation among organisms, but for any particular species the chromosomes remain constant in size, shape, and number. Chromosomes were recognized as early as the mid-nineteenth century and were thought to number 48. It was not until 1956 that the true number in man was determined to be 46 and to consist of 23 pairs of homologous chromosomes—22 pairs of *autosomes* and 1 pair of *sex chromosomes.* Both sex chromosomes are alike in the female (XX) but are morphologically different in the male (XY). This number (46) applies to all somatic cells and is referred to as *diploid,* meaning double or containing two sets. The gametes, or germ cells, contain one half the total number, or a single set referred to as *haploid* or *monoploid.*

Morphologically the chromosomes appear as dark-staining rodlike structures in the cell nucleus and are most easily seen under the light microscope during the

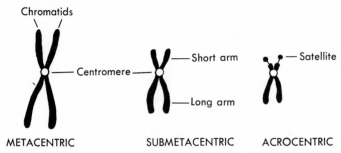

Fig. 1-3. The three classes of chromosomes and their major characteristics.

metaphase stage of cell division. At this stage the chromosomes are partially divided, with two longitudinal halves *(chromatids)* united at a constricted, pale or unstained area, the *centromere*. As cell division progresses, the two chromatids separate and each chromatid becomes part of a newly formed cell. The chromosomes, as they are usually depicted in photographs and diagrams, are actually two chromatids held together by a centromere and are classified according to their length and the position of the centromere (Fig. 1-3). The three classes in man are: (1) *metacentric*, those with a centrally located centromere and arms of equal length, which gives the X appearance; (2) *submetacentric*, those with the centromere nearer one end than the other and the arms of unequal length; and (3) *acrocentric*, those with a terminally located centromere, which gives the characteristic Y or wishbone appearance. On some acrocentric chromosomes a secondary constriction occurs in the distal end to form what is termed a *satellite*. Satellites are sometimes useful for identification and are significant in relation to some abnormalities.

Chromosome analysis

In order to be observed, chromosomes must be prepared and stained during the period when they are dividing. Not all of the various types of cells in the body are suitable; for example, nerve cells do not divide. The most commonly used cells are those obtained from bone marrow, skin, facia, or peripheral blood. All grow well in culture, but since they are easy to obtain, leukocytes are most frequently used. The leukocytes are separated from the other blood constituents and stimulated to grow in culture tubes at body temperature for approximately three days, during which time the cells divide. At the end of three days a substance, usually colchicine, is added that arrests cell division at the stage where the chromosomes are best visualized. The cells are then placed in a hypotonic solution. This causes them to swell and thereby spreads the chromosomes, which would otherwise remain clumped together. They are fixed, spread on a slide, stained, and photographed by a camera attachment on a high-power microscope.

After the photograph is enlarged the individual chromosomes are cut out and arranged according to a standard classification system, which was agreed upon by a group of researchers at a conference held in Denver in 1960 and revised at a con-

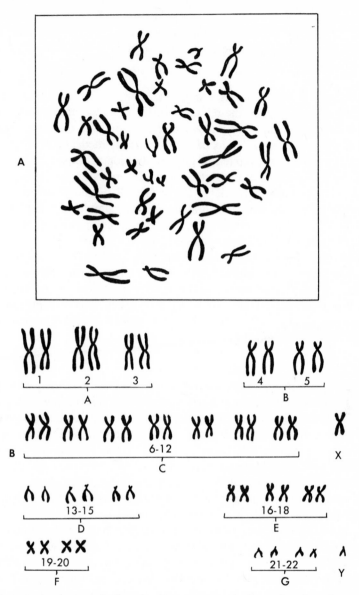

Fig. 1-4. A metaphase spread of male chromosomes. **A,** Example of a photomicrograph. **B,** Chromosomes arranged in a karyotype.

ference in Chicago in 1966. The paired chromosomes are positioned in order of decreasing length, assigned a number, and separated into seven major groups (designated A through G) composed of morphologically similar pairs. Most individual chromosomes cannot be identified with certainty but can readily be assigned to a group. This is essentially true with regard to number 6 through 12, the C group. The sex chromosomes, still designated X and Y, are either individually placed with the group that they resemble (the X chromosome with C group and the Y chromosome with G group) or paired to form an additional group. This systematic arrangement of chromosomes is termed a *karyotype* (Fig. 1-4).

CELL DIVISION

Every individual begins life as a single cell, the *zygote.* By the time he reaches adulthood his cells number in the billions. During this time numerous changes have taken place. The cells have not only increased in number but have differentiated into a variety of forms in order to perform the diversified functions of a complex organism. All the processes related to growth, development, and differentiation are made possible by somatic cell division, *mitosis.* In order for the organism to reproduce another of its kind, a specialized type of cell division, *meiosis,* is required during gamete formation. Fundamental to an understanding of heritable traits is a knowledge and appreciation of these basic means of cell reproduction.

Each cell division is both preceded and followed by an *interphase* stage. Interphase is the term used to describe a nondividing cell. (The formerly used term "resting stage" is no longer appropriate.) During this stage the nucleus, surrounded by a nuclear membrane, presents a granular appearance due to *nuclear chromatin.* Single chromosomes cannot be distinguished from one another and there appears to be no cellular activity, yet the cells are very actively engaged in the processes of protein synthesis and duplication of genetic material in preparation for division.

Mitosis

The process of mitosis describes the behavior of chromosomes during somatic cell division. Mitosis is an *equational division* in which cell components, particularly the genetic material, that were duplicated earlier are distributed in equal amounts to two daughter cells so that each possesses the same kind and number of chromosomes found in the mother cell. During this process the diploid set of chromosomes of the mother cell splits longitudinally and one half goes to each daughter cell, both of which will also be diploid.

Cell division is a continuous process, but for convenience in describing chromosome activity it is divided into four phases (Fig. 1-5).

Visible mitosis begins with *prophase.* The nuclear chromatin condenses and begins to take on a threadlike configuration. (The term mitosis is derived from the Greek word that means "the condition of making threads.") These threads continue to shorten and thicken until chromosomes become clearly visible and begin to migrate toward the center of the cell. At the same time, other events are taking place within the cell. A cytoplasmic structure, the *centrosome,* divides and the halves move to opposite ends of the cell. The nuclear membrane disintegrates, and the spindle fibers begin to form connections between the split centrosome.

By *metaphase* the nuclear membrane has disappeared. The clearly visible chromosomes, compressed to their shortest, thickest state, have aligned themselves in the center of the cell.

During *anaphase* the chromatids separate at the centromere and are drawn to opposite poles by the spindle fibers to which they are attached at the centromere.

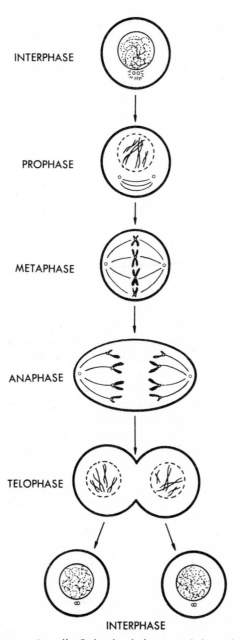

INTERPHASE

PROPHASE

METAPHASE

ANAPHASE

TELOPHASE

INTERPHASE

Fig. 1-5. Mitosis in a somatic cell. Only the behavior of four chromosomes is shown.

As the fibers contract and shorten, the centrosome appears to be pulling the arms of the chromosome toward the poles. They are now called *daughter chromosomes* or *daughter chromatids.*

The last stage of mitosis, *telophase,* begins when the daughter chromatids have completed their polar migration. They gradually become more diffuse and disperse, and at the same time a nuclear membrane forms around each.

The process of division of the cytoplasm, or *cytokinesis,* completes the cell division. The chromosomes take on a granular appearance, and the two daughter nuclei enter interphase. There is now an exact and equal distribution of genetic material in the two daughter cells. A diploid number of paired chromosomes has been produced for each new cell.

Meiosis

The process of mitosis, although a splendid method for duplicating somatic cells, would be totally unsatisfactory for germ-cell duplication. A full complement of chromosomes would, when combined with a similar set, increase from 46 to 1,472 in five generations. The process of meiosis (derived from the Greek word that means "to make smaller") results in cells whose nuclei contain half the chromosome complement of the parent cell so that when the gametes unite at fertilization, the original chromosome number is restored. This process is accomplished by two successive cell divisions during which one cell with a diploid (N = 46) set of chromosomes becomes four cells, each with a haploid (N = 23) set of chromosomes. Meiosis occurs only during germ-cell formation, or *gametogenesis,* in the testes *(spermatogenesis)* and ovaries *(oogenesis)* of the adult male and female.

As in mitosis, the meiotic chromosome divisions consist of four phases. The general aspect of meiosis will be considered, followed by the genetically significant features associated with sperm and ova formation (Fig. 1-6).

Meiosis I. In the first meiotic division the prophase is very long. Instead of migrating singly to the center of the cell, the chromosomes arrange themselves very closely in homologous pairs. This pairing, called *synapsis,* involves the entire length of the members so that each is matched centromere-to-centromere and gene-to-gene. The four chromatids are so intimately associated at this stage that the figure is sometimes referred to as a *tetrad.* It is at this point that an exchange of parts between chromatids, or *crossing over,* takes place. The significance of this will be explained later.

During metaphase I each spindle fiber attaches to a separate chromosome, rather than spindle fibers forming on either side of a centromere to split and pull apart the chromatids. At anaphase each intact member of a chromosome pair migrates to an opposite end of the dividing cell. The distribution of the maternally and paternally derived chromosomes is entirely random.

Anaphase I and telophase I proceed as in mitosis, but the end result is two cells with 23 chromosomes each and the sister chromatids still attached to one another at the centromere. With the formation of nuclear membrane and a cell wall, meiosis *(reduction division)* has taken place, resulting in two haploid sets of chromosomes.

Meiosis II. The second meiotic division is an equational division similar to

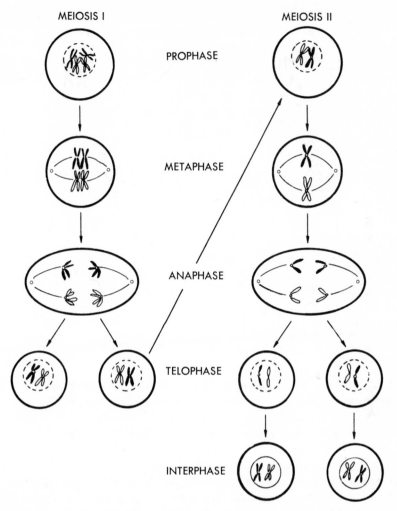

Fig. 1-6. Meiosis in a germ cell indicating the behavior of two pairs of homologous chromosomes.

mitosis. In each cell produced by the first division the chromatids of the unpaired chromosomes separate to form two haploid cells. The result is four cells with a haploid set of chromosomes. The complex nature of meiosis makes it more susceptible than mitosis to mechanical error. Most pathologic conditions that are due to chromosome abnormalities can be attributed to faulty division.

Spermatogenesis. The process of meiosis in the male gametes is a continuous process that begins about the time of puberty and continues until senescence. As they enter prophase I the diploid male germ cells (spermatogonia) are termed *primary spermatocytes.* These proceed through meiosis to form two haploid cells, the *secondary spermatocytes,* which in turn divide during meiosis II to yield four spermatids that each develop into a sperm. Throughout both divisions the cytoplasm

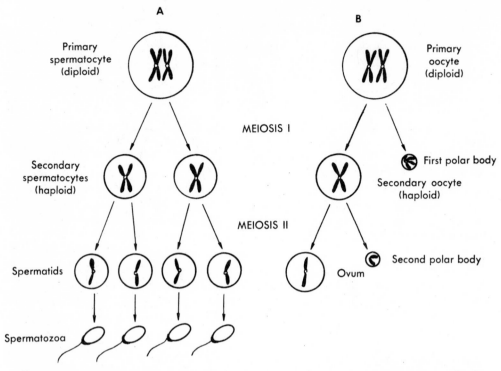

Fig. 1-7. Gametogenesis in male and female germ cells. **A,** Spermatogenesis results in four gametes (spermatozoa). **B,** Oogenesis results in only one gamete (ovum).

and the chromosome complement are evenly distributed to the resultant cells (Fig. 1-7, *A*).

Another significant feature of sperm formation is the division of the sex chromosomes. The male cells contain both an X and a Y chromosome, which, although morphologically dissimilar are homologous in pairing. During the first meiotic division the X and Y chromosomes go to opposite poles, each to produce two different types of cells in the second meiotic division (heterogametic). Consequently, there are two gametes with an X chromosome and two gametes with a Y chromosome produced from one primary spermatocytes.

Oogenesis. The process of meiosis in the female is not continuous. Oogenesis in the ovary begins during intrauterine life and proceeds to prophase I but remains at that stage until puberty. By the fifth to sixth intrauterine month the female has a full complement of *primary oocytes* that she retains throughout life. Each month, beginning at puberty and extending through menopause, one or two of these primary oocytes complete the meiotic cycle to produce a functional ovum or ova (Fig. 1-7, *B*).

The mature primary oocyte proceeds through the first meiotic division to form two daughter cells. Although there is equal division of the nucleus, there is unequal division of the cytoplasm. The result is one *secondary oocyte* containing the bulk

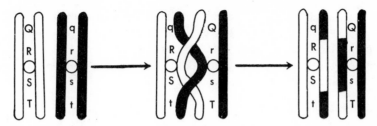

Fig. 1-8. Crossover in one pair of homologous chromosomes during meiosis.

of the cytoplasm and a smaller *first polar body* that is nonfunctional and soon disintegrates. Both proceed through meiosis II, which usually does not occur unless fertilization takes place. The secondary oocyte divides, again unequally, to produce an ovum and a nonfunctional *second polar body*. The cytoplasm retained by the oocyte seems to be essential to the embryo during early stages of development. It contains nourishment and the chemical messages necessary for development and differentiation of the zygote. There is speculation regarding transmission of some heritable characteristics through human maternal cytoplasm, but it has not been conclusively demonstrated.

The sex chromosomes in the oocytes are both X. Unlike the sperm, a mature ovum will contain only one type of sex chromosome (homogametic).

Linkage and crossover

Linkage is a term used to describe a situation in which two or more traits seem to occur together in an individual. When genes for these traits are situated on the same chromosome they travel together during cell division, and thus the individual will display both characteristics. Linkage occurs in all chromosomes, but in human beings specific linkage has been most easily established in relation to conditions associated with the X chromosome. The best example is the frequency with which colorblindness and hemophilia occur together in the same individual. Linkage is most apt to occur if the particular genes are situated very close together on the same chromosome. In this case the genes will remain together during the phenomenon of *crossover*.

Crossing over is a special feature of chromosomes during meiosis and accounts for the even wider variation of traits seen in individuals than would be explained by simple chromosome segregation. The mechanism of crossover takes place during prophase I when the paired chromosomes line up very close together (synapsis). The homologous chromosome pairs are so close together that there are one or more points of contact between their chromatids. At each point of contact, called a *chiasma,* there is a physical exchange of genetic material between the members of each pair. As a result, the newly formed chromosomes contain new combinations of maternal and paternal genes: each will contain genes from both the father and the mother (Fig. 1-8).

Crossing over is more apt to occur between genes situated far apart on a chromosome. By observing the frequency with which crossover occurs between genes on a given chromosome, researchers have been able to assess the relative distances between these genes. This serves as a basis for *gene mapping,* that is, identifying the location of genes that control specific phenotypic characteristics. At present only some of those genes borne on the X chromosome have been assigned to specific places with any degree of certainty.

CHEMICAL BASIS OF HEREDITY

The physical action and manifestations of the gene were the primary concern of geneticists for many years. All agreed that it was the basic unit of heredity— a discrete hereditary determiner—but the chemical nature of heredity was in doubt. Recently it has been shown that the genes are responsible for formation of specific proteins. It is now well understood that physical characteristics are dependent upon chemical reactions at the cellular level and, although it becomes increasingly complex, the one gene–one enzyme concept is inherently accurate: every metabolic process is regulated by an enzyme, which is controlled by a specific gene. Therefore, the most useful definition of a gene is one that describes its function. A gene

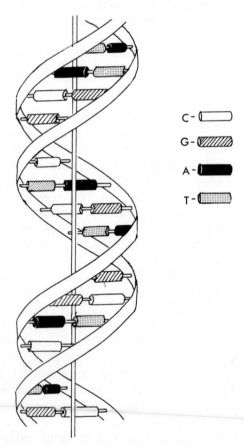

Fig. 1-9. The DNA molecule.

is a discrete section of a chromosome that contains the information necessary to produce an enzyme that will control a specific chemical reaction or group of reactions.

DNA

Although proteins had been considered the most basic cell compound, it is now known that genetic information is carried in molecules of *deoxyribonucleic acid* (DNA). Hereditary material must be able to duplicate itself and must provide for protein synthesis. The structure of DNA enables it to fulfill both of these functions.

Structure of DNA. The DNA molecule is composed of long chains of smaller molecules called *nucleotides*. These nucleotides are composed of a sugar molecule, a phosphate molecule, and a nitrogenous base. Each nucleotide contains the same sugar base (deoxyribose) and the same phosphate base (phosphoric acid) but differs in the nitrogenous base. The nitrogenous bases consist of the purines *adenine* and *guanine* and the pyrimidines *cytosine* and *thymine*. It is these nitrogenous bases that provide versatility and variation in genes and protein synthesis.

Because of the Nobel Prize–winning accomplishments of Watson and Crick, the DNA molecule is known to be composed of not one but two nucleotide chains loosely attached at their bases and coiled around each other to form a double-stranded helix (Fig. 1-9). This formation is sometimes described as a twisted ladder or spiral staircase.* Alternating sugar and phosphate molecules provide the "backbone" or supports. Attached to the sugar molecules, the nitrogenous base of one chain of nucleotides is bound to the base of the other by hydrogen bonds to form the rungs of the ladder (Fig. 1-10). The most significant feature of this arrangement is the precise relationship maintained by the purine and pyrimidine bases. A purine in one chain always pairs with a pyrimidine in the other; more specifically, adenine (A) always binds with thymine (T) and cytosine (C) always binds with guanine (G).

> *Purine* binds to *pyrimidine*
> adenine—H—thymine
> guanine—H—cytosine

Therefore, if the base sequence in one chain is

A G A T C C G T A A A C

the complementary chain must contain the bases in the order

T C T A G G C A T T T G

Replication† of DNA. The Watson-Crick Model of DNA also provides a theoretical explanation of how the DNA is able to duplicate itself exactly in the interphase stage between cell divisions; this theory has received solid support from

---

*This does not mean that the DNA helix is actually a chromosome. Chromosomes are very much larger than DNA molecules and also contain, in most organisms, large amounts of protein. Exactly how the DNA forms the chromosome is still not understood.
†The terms duplication and replication are synonymous and are used interchangeably.

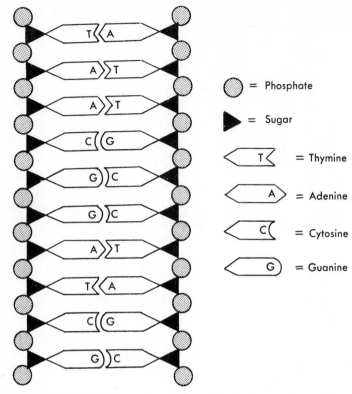

Fig. 1-10. Base pairing of nucleotides in a DNA molecule. A attaches to T, and C attaches to G.

other investigators. When DNA replicates, the original strands first separate at the hydrogen bonds to form two separate single strands (a process that has been likened to that of unzipping a zipper). As the hydrogen bonds break, the strands "unzip," and bases from a pool of free nucleotides in the cell nucleus attach themselves to the appropriate bases in the chain to form a complementary chain. The result is that two new chains are constructed along the original chains to form two new helixes, each chemically identical to the helix from which it was derived (Fig. 1-11). When cells divide in this way the genetic information is preserved and transmitted unchanged to the daughter cells.

RNA

Whereas DNA is a chemical constituent of all chromosomes and found primarily in the cell nucleus, a sister compound, *ribonucleic acid* (RNA), is found in both the nucleus and the cytoplasm. The molecular structure of RNA is similar to that of DNA except that RNA contains a slightly different sugar, *ribose,* and in RNA one pyrimidine, thymine, is replaced by a structurally similar nucleotide, *uracil* (U). Also, RNA usually exists as a single strand rather than a double

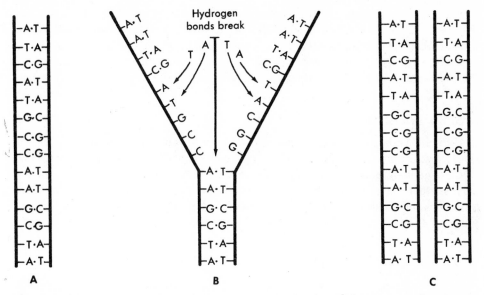

Fig. 1-11. DNA replication. **A,** Portion of one DNA molecule. **B,** The two strands separate at the hydrogen bonds, and free bases attach to the appropriate base in the original strand. **C,** Two new DNA molecules, each identical to the original molecule.

strand. The significant role played by RNA in cellular processes is seen in protein synthesis.

THE GENETIC CODE

Very simply, the genetic code is the language of the genetic messages. When researchers assumed the task of cracking the genetic code, it was obvious that individual nucleotides could not code for the 20 known amino acids; therefore, a combination of symbols must represent an amino acid. A doublet, or two-symbol code, allows only 16 possible combinations ($4^2 = 16$) whereas a triplet, or three-symbol code, provides a possible 84 combinations ($4^3 = 64$)—more than enough to code for the 20 amino acids. A triplet, or sequence of three adjacent nucleotides (bases), in the DNA or RNA molecule that codes for an amino acid is called a *codon.* Research has substantiated this triplet code as the basic genetic language, and the specific combination of bases has been identified for each amino acid (Appendix A). Since the message is formed by mRNA (messenger RNA; p. 21), the codes are usually designated by the RNA bases C, G, A, and U, and more than one triplet can code for the same amino acid. For example, the code for phenylalanine has been determined to be UUU or UUC; the codes for arginine are CCU, CGC, CGA, or CGG. It has been questioned whether the triplet code is overlapping. For example, if the base sequence is UAAGUGCACU, then the possible triplets might be UAA, AAG, UGU, GUG, UGC, GCA, CAC, or ACU. The sequence of bases in one of the two polypeptide chains of the DNA molecule dictates the order in which the amino acids are incorporated into the chain. Polypeptide chains combine to form enzymes and structural proteins. If nucleotides were attached in a random fashion the result would be erratic and inconsistent. Experiments have shown that during protein synthesis, attachment starts at one end of the RNA strand and proceeds in an orderly sequence to the other end. There is also evidence to indicate that there are three terminator codons that signal the end of polypeptide chain synthesis.

PROTEIN SYNTHESIS

Protein synthesis takes place in the cytoplasm at intracellular structures termed ribosomes, a name derived from their high content of RNA with its ribose. Since DNA containing genetic information is primarily fixed in the chromosomes of the nucleus, there must be a link between the chromosomes and the ribosomes. RNA forms this link. The process of protein synthesis begins in the nucleus during interphase of the cell cycle (Fig. 1-12). The nucleotide sequence in the DNA of the

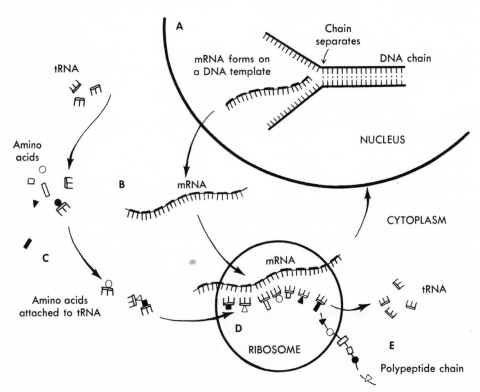

Fig. 1-12. Protein synthesis (simplified). **A,** DNA molecule separates. RNA forms a template from one DNA strand. **B,** mRNA template migrates to a ribosome. **C,** tRNA picks up amino acids from cytoplasm and carries them to the ribosomes. **D,** tRNA with its amino acid finds a matching triplet on mRNA (now ribosomal RNA) to form a polypeptide chain. **E,** Polypeptide chain formed and released from tRNA; tRNA released in cytoplasm; mRNA returns to nucleus for recoding.

chromosomes forms a pattern or template upon which RNA molecules are built in a manner similar to DNA duplication. The term for this process is *transcription.* After the RNA has copied the message, it detaches from the DNA. With the message intact, these molecules of RNA then migrate to the cytoplasm where they, in turn, form a template for the formation of proteins as directed by the original DNA. This has been termed *messenger RNA* (mRNA), although it is in fact a message rather than a messenger.

The mRNA template attaches to a ribosome where the specified protein will be constructed. The amino acids from which proteins will be formed are disseminated throughout the cytoplasm; consequently, there must be some means of attracting the proper amino acid to the correct site of the RNA template. *Transfer RNA* (tRNA)* fulfills this function. It is believed that for each amino acid there is a specific tRNA. The tRNA with its specific amino acid finds the point of attachment on the mRNA template where it and no other can fit. For example, an amino acid specified by the DNA bases GCC can fit only into a space coded by the mRNA CGG, and an amino acid specified by the DNA bases AAA can fit only the code UUU (remember that T has been replaced by U in RNA). In this way the amino acids are aligned in the correct sequence to form the specified protein, for example, phenylalanine or valine. The term for this process is *translation.* Protein synthesis can be diagrammed:

$$\text{DNA} \xrightarrow{\text{transcription}} \text{RNA} \xrightarrow{\text{translation}} \text{PROTEIN}$$

The process of protein synthesis can be briefly stated: DNA makes RNA and RNA makes protein.[1]

Control of gene action

Not all genes are responsible for determining the amino acid sequence of proteins; some perform a controlling role. All the genes are not active at the same time in any cell; therefore, the control mechanism functions to allow some genes to act and to keep others inactive. Genes, then, are classified into two basic types: (1) *structural genes,* which are those directly concerned with the production of mRNA and thus the proteins and (2) *control genes,* which are (a) *operator genes* that initiate the activities of a series of structural genes, and (b) *regulator genes* whose function is to control the operator genes. The operator gene initiates the formation of polypeptides by the specific group of structural genes under its control. Each operator gene is in turn controlled by a *repressor substance* produced by a regulator gene. This repressor, acting through the cytoplasm, inhibits protein synthesis by action on the operator gene. In the absence of the repressor substance the operator is said to be *derepressed* and, thus, free to allow protein synthesis to proceed (Fig. 1-13). Further research regarding the activities of this gene may have significant medical implications, especially regarding excesses and deficiencies of various proteins and the importance of external influences that might alter its activities.

---

*Some researchers prefer to use the term *soluble RNA,* or sRNA, to describe this form of ribonucleic acid.

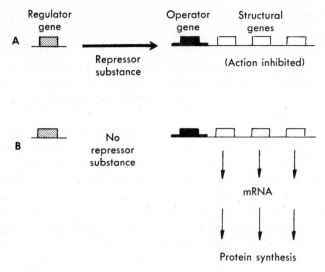

Fig. 1-13. The action of regulator and operator genes.

GENE MUTATION

A mutation is a change in genetic material. A mutation can be either spontaneous or induced and, although a rare occurrence, has important genetic significance. Some gene mutations are the result of changes in chromosome structure, such as deletions or additions of chromosome material or unequal crossover. The most frequently occurring mutations are the result of chemical changes. If the sequence of bases in a DNA molecule is replaced or altered, the resulting protein will have totally different properties.

The alphabet, word, and sentence analogy illustrates the effect of gene mutation.[2] If each base combination represents a letter of the alphabet and amino acids the words, the protein would be a sentence. If only one letter in the sentence "He was a chief" were replaced to read "He was a thief," the sentence conveys an entirely different meaning. When a letter or group of letters are merely rearranged, "He was a thief" becomes "He saw a thief," and again the meaning is quite different.

Genes, too, can change or rearrange. A single gene can mutate to produce an entirely different protein. When a chromosome breaks and reattaches in reverse sequence, the consequence is an altered protein. Genes are usually stable. When a gene changes, the mutant gene remains equally stable and is transmitted to future generations. A good example of a gene mutation is in relation to the hemoglobin molecule. When one of the amino acids (glutamic acid) in normal hemoglobin A is replaced by another amino acid (valine) to form hemoglobin S, the result is sickle cell anemia (p. 63). Mutation has meaning for individuals and for populations as a whole (p. 170). Its importance will be observed in relation to various aspects of heredity.

REFERENCES

1. Crispins, C. G., Jr.: Essentials of medical genetics, New York, 1971, Harper & Row, Publishers, Chapter 4.
2. Goldstein, P.: Genetics is easy, ed. 4, New York, 1967, The Viking Press, Inc., p. 221.
3. McKusick, V. A.: Human genetics, ed. 2, Englewood Cliffs, N. J., 1969, Prentice-Hall, Inc., p. xi.

GENERAL REFERENCES

Chicago conference: Standardization in human cytogenetics. Birth defects, original article series **2:**2, New York, 1966, The National Foundation.

Crick, F. H. C.: The genetic code, Sci. Am. **207:**66, 1962, part one; **215:**55, 1966, part three.

Denver report: A proposed standard system of nomenclature of human miotic chromosomes, Lancet **1:**1063, 1960.

Eggen, R. R.: Chromosome diagnostics in clinical medicine, Springfield, Ill., 1965, Charles C Thomas, Publisher.

Fraser, A.: Heredity, genes, and chromosomes, New York, 1966, McGraw-Hill Book Co.

Levine, L.: Biology of the gene, ed. 2, St. Louis, 1973, The C. V. Mosby Co.

Nirenberg, M. E.: The genetic code, Sci. Am. **208:**80, 1963, part two.

Paris conference (1971): Standardization in human cytogenetics. Birth defects, original article series, **8:**7, New York, 1972, The National Foundation.

Wagner, R. P., and Mitchell, H. K.: Genetics and metabolism, ed. 2, New York, 1964, John Wiley & Sons, Inc.

Watson, J. D., and Crick, F. H. C.: The structure of DNA, Cold Springs Harbor Symposia, 1953, vol. 18. Reprinted in Levine, L.: Papers on genetics, St. Louis, 1971, The C. V. Mosby Co.

2

Gene transmission in families

Disorders for which a simple, definite inheritance pattern can be identified are rare individually, but collectively they constitute a sizable portion of conditions that concern health workers. The incidence of genetically determined disorders with serious consequences varies from approximately 1 in 2,000 for the more frequently occurring diseases to only a handful of reported cases for some very rare diseases. The variety of disorders, however, numbers in the hundreds. McKusick[1] has compiled a volume in which he describes over 1,800 disorders known to be transmitted by simple mendelian inheritance patterns.

Heritable disorders can involve any system in the body. Structural defects are readily apparent in a newborn infant, and some of the inborn errors of metabolism can be detected within a few days. Many disorders, such as the muscular dystrophies, do not become evident until early childhood or even adulthood in some varieties. Some defects can be of such minor importance that they have little or no effect upon survival; others are so serious as to be incompatible with life. An extra digit or premature baldness does not interfere with survival, and those afflicted can carry on a relatively normal existence. More serious conditions such as cystic fibrosis and hemophilia are a threat to survival and interfere with the activities of the affected individuals and their families as well. A disease or disorder that can be transmitted from one generation to the next is termed *hereditary* or *heritable* and may or may not be apparent at birth. A *congenital* disorder is one that is present at birth and can be due to either hereditary or environmental causes.

When researchers attempt to study heritable characteristics in animals they can isolate a trait through manipulation and regulation of breeding within the species. It is, of course, impossible to conduct such controlled experiments with human beings. However, a great deal of knowledge regarding the way that heritable disorders are transmitted has been acquired through the study of families in which the disorder is manifest. Conditions due to a single gene have been identified by applying this information to genetic theory. Most are relatively rare conditions, and the more rare the disorder the easier it is to observe the inheritance pattern. The more common but less clear-cut conditions (those in which no single gene can be identified) present a complex problem and will be discussed in Chapter 6.

PRINCIPLES OF INHERITANCE

The principles of inheritance form the foundation upon which rests the science of genetics; therefore, a knowledge of these fundamental laws is essential to an understanding of heritable disorders. Basically the same today as when first described by Gregor Mendel in the mid-nineteenth century, these laws describe gene activity during gametic formation and fertilization. His observations are still designated as *Mendel's laws* and the statistically predictable results as *mendelian ratios.* Many of the rare traits seen in man are distributed in families in characteristic patterns according to the laws of Mendel.

Principle of dominance

Each individual possesses two genes that determine any given trait, but not all genes operate with equal vigor. When a chromosome pair contains dissimilar genes at a specific locus, a competition will exist between them for expression in the individual. As a result of this competition, one may mask or conceal the other. The characteristic (trait) that is manifest in the individual (and the gene that produces the effect) is referred to as *dominant;* that which is hidden and not manifest is *recessive.* In the example of genes for eye color, if the genotype contains a gene each for blue and brown eye color and brown eye color is displayed in the phenotype, the gene for brown eye color is dominant and that for blue eye color is recessive. The dominant characteristic is always displayed in the phenotype, and it is usually impossible to tell if the genotype is homozygous for the trait. To be expressed, a recessive characteristic must always be homozygous. An exception to this rule is seen when only one gene is present and therefore its effects are exhibited in the individual (for example, the X-linked traits).

In regard to physical disabilities and disorders, a dominant disorder is one in which the individual who carries the gene (either heterozygous or homozygous) is clinically affected; a recessive disorder is one in which the clinically affected individual is homozygous.

Principle of segregation

The *law of segregation,* sometimes called Mendel's first law, describes the separation of alleles during gamete formation and their recombination during fertilization. During reduction division (meiosis) the paired chromosomes bearing traits derived from each parent are separated to form two gametes. The genes remain unchanged during this process so that each gene is essentially the same as it had been through countless generations.

It would be unrealistic to use human examples to illustrate the law of segre-

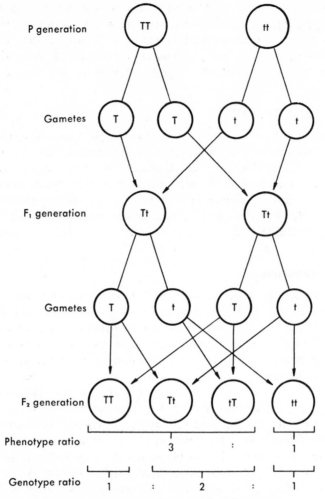

Fig. 2-1. Segregation of a pair of allelic genes.

gation; therefore, Mendel's garden pea experiment will be briefly outlined. One of the numerous traits that Mendel observed and analyzed in his experiments was plant height. When tall peas were mated with short peas in the parental (P) generation, all the offspring in the first filial (F_1) generation resembled only the tall parent—they all expressed the dominant trait (Fig. 2-1). T is used to indicate the dominant (tall) allele, and t represents the recessive (short) allele. When the plants from the F_1 generation were self-pollinated, the resulting offspring in the F_2 generation expressed both dominant and recessive traits (tall and short) in the ratio of 3:1. When the recessive plants were self-pollinated, all the progeny exhibited the same trait: all were short. However, when the plants exhibiting the dominant trait were self-pollinated, they displayed both traits. Two thirds of the plants yielded offspring in the ratio 3:1 as in the F_2 generation, and

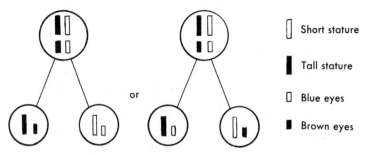

Fig. 2-2. Independent assortment of two pairs of allelic genes.

the remaining one third produced only tall plants. In the F_2 generation, then, although the plants displayed a phenotype ratio of 3:1, the genotype ratio was 1:2:1. That is, one fourth of the plants were *TT*, or homozygous dominant; half were *Tt*, or heterozygous, and displayed the dominant trait; and one fourth were *tt*, or homozygous recessive. Each gene segregates in pure form unaffected by association with its contrasting allele, and chance alone determines which gene will travel to which gamate.

Genetic counselors apply the law of segregation in predicting the probability of a homozygous recessive gene combination occurring in a child from a mating between heterozygotes and often to assess the genotype of parents (if there are a sufficient number of children).

Principle of independent assortment

The basic premise of *independent assortment,* or Mendel's second law, is that when characteristics displayed in an individual have alleles at two or more loci, each is distributed in the gametes in random fashion independent of the others. For example, if a parent cell contains genes for blue and brown eye color and genes for ability or inability to taste phenylthiocarbamide (PTC), a drug with a bitter taste detected by persons with the dominant gene but unable to be detected by persons homozygous for the recessive allele (p. 163), they might segregate in any of the following combinations: taster/brown eyes, nontaster/ brown eyes, taster/blue eyes, or nontaster/blue eyes (Fig. 2-2). Thus, with only two traits a germ cell could produce four combinations of traits in the gametes. The number of possible combinations using the thousands of genes contained on only one pair of the 46 chromosomes is staggering.

Fig. 2-3 illustrates the use of a Punnett square to compute the variety of trait combinations in offspring that could result from mating of parents who are heterozygous for these same two traits. Where tasters *(T)* and brown eyes *(B)* are dominant and nontasters *(t)* and blue eyes *(b)* are recessive, the phenotypic ratio in offspring of such a mating would be 9:3:3:1. That is, nine would be brown-eyed/tasters, three would be blue-eyed/tasters, three would be brown-eyed/ nontasters, and one would be a blue-eyed/nontaster. The genotype ratio is much more elaborate. One of the nine brown-eyed/taster offspring is *T/T, B/B;* two are *T/T, B/b;* four are *T/t, B/b;* and two are *T/t, B/B* (Fig. 2-3).

Gametes	TB	Tb	tb	tB
TB	T/T B/B	T/T B/b	T/t B/b	T/t B/B
Tb	T/T B/b	T/T b/b	T/t b/b	T/t B/b
tb	T/t B/b	T/t b/b	t/t b/b	t/t B/b
tB	T/t B/B	T/t B/b	t/t B/b	t/t B/B

Fig. 2-3. A variety of mating types and expected proportion of offspring for two pairs of allelic genes **T, t, B,** and **b.** Offspring in shaded and white squares have the same phenotype.

		Parent A/A	
	Gametes	A	A
Parent A/A	A	AA	AA
	A	AA	AA

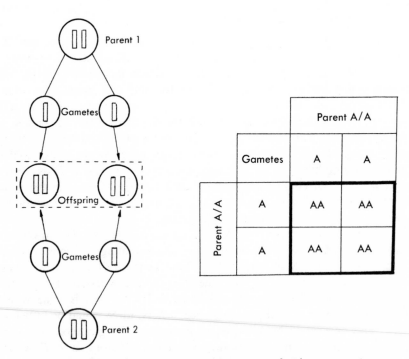

Fig. 2-4. Mating between parents homozygous for the same trait.

Inheritance models

There are several possible combinations of dominant and recessive genes that segregate or assort during gamete formation and combine during fertilization to produce offspring in the predictable mendelian ratios. Figs. 2-4 through 2-7 are models of the various segregation and combination patterns for a single set of allelic genes. A diagram of genes during meiotic division and the Punnett square are useful tools for determining the possible outcome of any combination of traits, and both are used in these examples. Upper case letters indicate the dominant allele and lower case the recessive.

When the parents are homozygous for a given trait, either dominant or recessive, all the offspring will display the trait and in turn pass it on to their offspring (Fig. 2-4). An individual with a trait known to be recessive is always homozygous for that trait.

When parents are each homozygous for dissimilar traits, the phenotype of the offspring will all resemble the dominant parent (Fig. 2-5). However, the genotype (gene composition) of the offspring will be heterozygous and contain the genes for both traits seen in the parents.

If parents are heterozygous for a trait, both will display the dominant trait. Their offspring will display both the dominant and recessive traits in a 3:1 ratio— three with the dominant trait and one with the recessive trait. However, genotypically, one offspring will be homozygous for the dominant trait, two will have

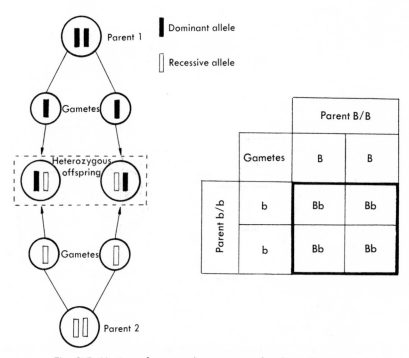

Fig. 2-5. Mating of parents homozygous for dissimilar traits.

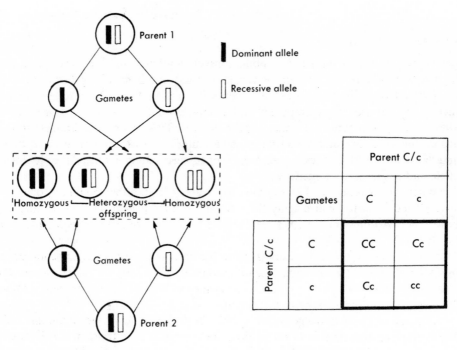

Fig. 2-6. Mating of parents heterozygous for the same traits.

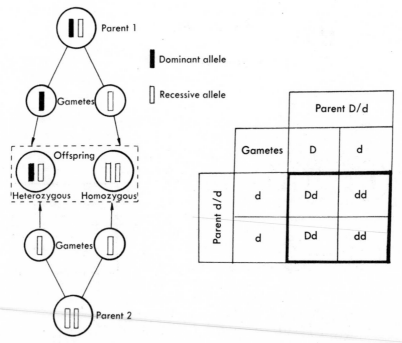

Fig. 2-7. Mating of a parent homozygous for a recessive trait and a parent heterozygous for a recessive trait.

heterozygous genotypes but display the dominant trait, and one will be homozygous for and display the recessive trait—a 1:2:1 ratio (Fig. 2-6). Disorders due to a recessive gene are usually transmitted in this manner.

Another possible mating is between a heterozygous parent who displays a dominant trait and a homozygous parent who displays the recessive trait. The predicted offspring will display the dominant and recessive characteristics in a 1:1 ratio. Both offspring who display the dominant phenotype will be genotypically heterozygous. The other two will display the recessive trait and will have a homozygous genotype (Fig. 2-7). This is the combination usually seen in disorders due to a dominant gene.

THE PEDIGREE CHART

The task of observing the distribution pattern of a specific trait in a family is facilitated by the construction of a family tree or *pedigree chart*. Data are summarized in a pedigree chart by use of symbols to indicate individuals, relationships, and significant details related to them. There are elaborations on some complex histories and by various investigators, but the basic and most frequently used symbols are outlined in Fig. 2-8. The construction of a pedigree begins with the affected person, who is referred to as the *proband, propositus* (or *proposita* if a female), or *index case* and is designated with an arrow. Marriages are represented by a bar, with males usually indicated first on the left. A double bar indicates a marriage between blood relatives, or a *consanguineous* marriage. Siblings are indicated by arabic numerals from left to right in order of birth. Generations are represented by roman numerals, the earliest at the top. Abbreviated pedigree charts will be used throughout this chapter to illustrate the various inheritance patterns for disorders due to mutant genes. For simplicity, normal spouses are usually omitted when a chart is used for illustration.

THE SINGLE MUTANT GENE

Many conditions in man for which a single mutant gene is directly responsible are distributed in families according to the basic principles described in the previous sections. The major inheritance patterns, based on these principles, govern the transmission of any set of genes, and most disorders due to a single gene that produces a large effect are readily recognized by the simple mendelian family patterns that they display. A trait determined by a gene on an autosome is referred to as an *autosomal trait* and may be dominant or recessive. A trait determined by a gene borne on one of the sex chromosomes is said to be *sex-linked* and also can be dominant or recessive.

Some generalizations have been observed regarding diseases and malformations due to a single gene. Disorders due to structural defects seem to be primarily the result of dominant genes, while a recessive inheritance pattern is seen in most of the inborn errors of metabolism. Dominant traits are seen more frequently and are usually less severe than recessive traits. This is probably due to the "double dose" effect. That is, recessive traits are only manifest when both genes are present, whereas the gene pairing in dominant disorders can be either

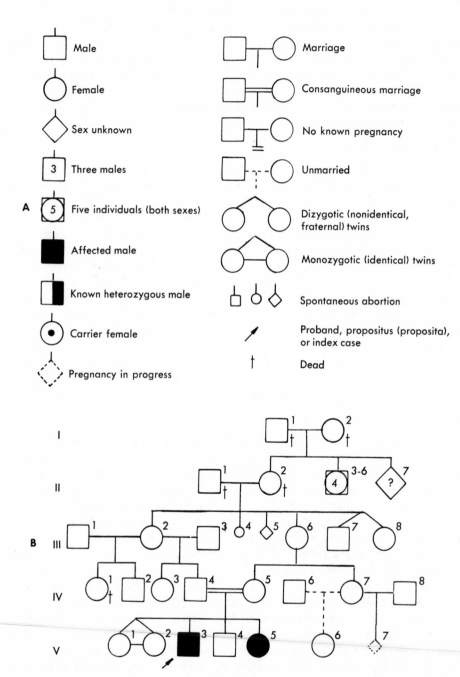

Fig. 2-8. A, common pedigree symbols and **B,** an example of a pedigree.

heterozygous or homozygous. The presence of the normal gene seems to have the effect of reducing the severity of a dominant disorder. Some dominant disorders when observed in the heterozygous state are relatively mild but when seen in the homozygous state can be so extreme as to be incompatible with life.

In genetics the term *lethal* is used to describe any condition that produces such a drastic effect that the zygote dies in utero and is subsequently aborted or the individual is unable to bear offspring. In each situation the genetic consequences are the same. A dominant disorder so severe as to render reproduction impossible will not be passed on to future generations. However, a recessive trait, even though it may be lethal or *semilethal* (the individual dies at an early age), can be transmitted unexpressed in the heterozygote through numerous generations.

AUTOSOMAL INHERITANCE

Since there are 44 autosomes and only 2 sex chromosomes, the majority of disorders are due to a single mutant gene on the autosomes. Autosomal conditions are displayed with equal frequency in both males and females.

Autosomal-dominant inheritance

Disorders due to a dominant gene on an autosome are those in which the gene for the abnormal trait dominates and the normal gene is recessive. The trait is expressed in both the homozygote and the heterozygote. Whenever the gene is present it is expressed in the phenotype; therefore, it can be traced through a number of generations. This is termed a *positive* family history (Fig. 2-9).

An individual affected with an autosomal-dominant disorder will have an affected parent unless the condition is the result of a fresh mutation, and the trait usually occurs in one parent only. It is extremely rare to see both parents with the same disorder. Half the sons and half the daughters of an affected parent will display the abnormality. Since the abnormal gene is located on only one of a pair

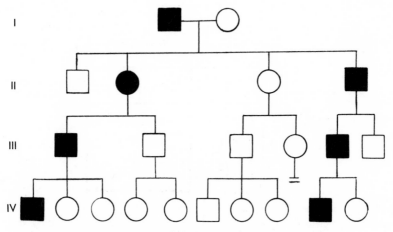

Fig. 2-9. An autosomal-dominant inheritance pattern.

of autosomes, any given offspring will have an equal chance to receive either the normal or the mutant gene.

The unlikely mating of two heterozygous affected parents will produce affected children in the ratio 3:1. For a parent to be homozygous would be merely hypothetical, since few autosomal-dominant traits have ever been observed in the homozygous state. In such a situation the offspring would all be affected.

Since the normal children of affected parents will not carry the abnormal gene, their offspring will be unaffected. Usually the first case in a family appears suddenly as the result of a fresh mutation and, depending on the degree of disability the condition imposes on the individual, will either die out or continue to be passed on through several generations. Some autosomal-dominant disorders are so severe that the affected individual may be infertile or die at an early age. In these cases, the disease becomes extinct in the family and is maintained in the population only by fresh mutations in other families. Other disorders, such as polydactyly or achondroplasia (dwarfism), have little effect on survival.

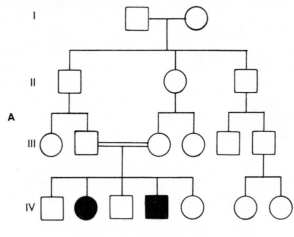

Fig. 2-10. An autosomal-recessive inheritance pattern. **A,** Pedigree chart. Note the consanguineous marriage in generation III. **B,** Punnett square illustrating the offspring ratio from the marriage of two heterozygous carriers of a recessive allele.

The survival of a dominant gene is also influenced by the age at which the disorder is manifested. The classic example of this is Huntington's chorea, a disabling and ultimately fatal disease of the brain characterized by disordered and involuntary muscular movements and accompanied by progressive mental deterioration. The average age of onset is 35 years of age, with the first signs usually appearing some time between the ages of 25 and 55. By the time signs of the disease appear, an affected individual can produce a sizable family. Occasionally a person who carries the gene for this disorder will die from other causes before there is any indication of its presence. This is one of the diseases for which relatives of affected persons most frequently seek genetic counseling.

Major characteristics of an autosomal-dominant trait can be summarized: (1) males and females are affected with equal frequency, (2) affected individuals will have an affected parent, (3) half the children of a heterozygous affected parent will be affected, (4) normal children of affected parents will have normal children, and (5) traits can be traced vertically through previous generations—a positive family history.

Autosomal-recessive inheritance

A normal gene will dominate and mask a gene for an autosomal-recessive trait; therefore, individuals who display an autosomal-recessive disorder will always be homozygous for that trait. For this reason most individuals who are heterozygous for a rare recessive disorder remain undetected in the population. Although there is variation among authorities, the estimate is that each person carries from 3 to 8 genes capable (when combined with a similar gene) of producing a lethal or semilethal disorder. This usually creates no problem because the probability of random mating of two persons carrying the same lethal gene is highly unlikely. However, the probability of marriage between heterozygotes is increased when the individuals are blood relatives; that is, when the marriages are consanguineous. This hazard has been recognized for centuries. Nearly every culture has had some taboo regarding marriage between relatives, and marriages between close relatives are illegal in many states.

Almost all the individuals with an autosomal-recessive disorder have normal parents, and there is usually no evidence of the trait in previous generations—a *negative* family history (Fig. 2-10). The disorder can result from either a new mutation or from the mating of parents who are both heterozygous for that particular abnormal gene. The disorder is only manifest when the gene is present in double dose; therefore, the affected individual must receive the abnormal gene from each parent. If one parent is affected and the other parent is homozygous normal, *none* of the offspring will display the disorder but *all* will be heterozygous carriers of the abnormal gene. Although improbable, if both parents are affected with the same recessive trait (homozygous) all their children will be affected.

Just as in autosomal-dominant conditions, there is no sex difference; the autosomal-recessive disorder occurs with equal frequency in males and females. Parents heterozygous for a recessive trait can expect that one fourth of their children will be affected. It is not uncommon for the trait to be manifest in one or more siblings (except half-sibs) and confined to that particular sibship. It is also possible that none of the offspring will be affected.

Even though the dominant allele seems to inhibit its deleterious effects, the evidence of a mutant recessive gene can usually be detected in the heterozygote once the basic defect has been determined. Because of an increasing number of inborn errors of metabolism, this is being carried out with suspected heterozygotes such as the siblings of an affected person. Methods for detecting the presence or absence of a recessive trait in the heterozygote are being developed for more and more of these disorders. Given sufficient time and more refined diagnostic tools, the screening for heterozygous carriers may become commonplace.

Major characteristics of an autosomal-recessive disorder can be summarized: (1) males and females are affected with equal frequency, (2) affected individuals will have unaffected parents who are heterozygous for the trait, (3) one fourth of the children of two unaffected heterozygous parents will be affected, (4) two affected parents will have affected children exclusively, (5) affected individuals married to unaffected individuals will have normal children (heterozygotes), and (6) there is usually no evidence of the trait in antecedent generations—a negative family history.

Intermediate inheritance

In some situations the degree to which a gene produces its effect does not always follow the dominant-recessive principles of single-gene effects. The feeling now seems to be that dominance and recessiveness are somewhat arbitrary concepts and that the dominant or recessive expression in the phenotype is greatly dependent upon the ability of diagnostic methods to identify the presence of the products of gene action (by definition a dominant gene is one that is expressed in the phenotype). As indicated earlier, often the presence of the recessive gene can be detected in the normal heterozygote, and in a dominant disorder the normal gene modifies the severe effect of a dominant gene. The gene for a disorder, either dominant or recessive, will create an extreme effect in the homozygous state (a double dose). There are some conditions in which this situation is readily apparent, and the term *intermediate inheritance* is sometimes applied to this pattern of gene transmission. It is somewhere in between dominant and recessive. The gene is partially expressed in the heterozygote but differs qualitatively and quantitatively in the homozygote. Sometimes this type of expression is described as *incompletely dominant* or *incompletely recessive*. The singing voice is a common illustration of this type of inheritance. The tenor and base in the male and the soprano and contralto in the female are homozygous; the baritone and mezzo soprano are heterozygous.

The inheritance pattern of an intermediate gene is the same as that for an autosomal-recessive trait since the severely affected individual is always homozygous and the parents are heterozygous carriers of the trait. An example of such a situation occurs with the gene for hemoglobin S, or sickle hemoglobin (Fig. 2-11). The phenotypic expression in individuals who are homozygous for the abnormal gene is the severe sickle cell anemia. Under conditions of reduced oxygen

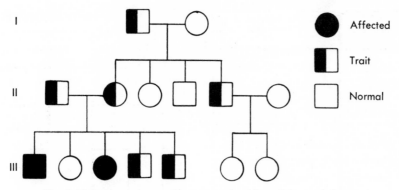

Fig. 2-11. Intermediate inheritance of sickle cell hemoglobin.

concentration the hemoglobin S causes a high percentage of red blood cells to change from round to a sickle shape, with a variety of serious consequences. In many parts of the world persons with sickle cell anemia do not survive to reproduce.

The blood of heterozygotes contains both types of hemoglobin, and they appear healthy and normal although their red blood cells can be caused to sickle in the laboratory when subjected to conditions of low oxygen tension. These heterozygous individuals are said to possess *sickle cell trait* and have a normal life expectancy. In some areas the trait seems to provide an advantage over the individual with normal hemoglobin (p. 173).

Similarly, thalassemia, a blood disease in which there is interference with the production of normal hemoglobin, is in most instances a relatively harmless condition in the heterozygote (thalassemia minor). But when exhibited in the homozygous state (thalassemia major, or Cooley's anemia) the affected persons usually do not survive to adulthood. This disease is primarily restricted to persons with Italian ancestry or, to a lesser extent, some other peoples of the Mediterranean area.

Codominance

Another situation in which both alleles of a gene pair are equally expressed in the heterozygote is codominance. An illustration of this is the ABO blood groups. These blood groups are present in the population as *multiple alleles;* that is, there are more than two genes at the gene loci that produces the variety in blood types. Consider the three most important of the blood group genes: A, B, and O. Instead of the usual possible genotypes (homozygous dominant, heterozygous, and homozygous recessive), there are six possible combinations from these three genes: A/A, A/O, B/B, B/O, A/B, and O/O. Persons with type AB blood demonstrate the effects of the genes for both antigen A and antigen B. Since neither is recessive to the other they are said to be *codominant*. The gene for type O is recessive to both A and B and is expressed only in the homozygous state. As a result, only the types A, B, AB, and O are seen in the phenotype. It

Phenotype	Genotypes	
Type A	AA	AO
Type B	BB	BO
Type O	OO	—
Type AB	AB	—

Fig. 2-12. Codominant inheritance of the ABO blood groups. A, Possible genotypes from the phenotypes A, B, AB, and O. B, All four phenotypes can be produced from mating of A/O and B/O genotypes.

is possible for parents, one heterozygous for type A and the other heterozygous for type B, to produce children of each phenotype (Fig. 2-12).

The MN blood group, although it consists of only two genes, M and N, are codominant and behave in the same manner as A and B. Blood group genetics, in addition to use in compatibility determination for transfusion, is of value to anthropologists and in legal situations of disputed paternity.

SEX-LINKED INHERITANCE

Genes on the sex chromosomes are transmitted according to the principles of heredity just as they are on the autosomes. However, sex chromosomes differ from autosomes morphologically and in gene complement. Therefore, in the transmission of some traits the inheritance patterns will vary according to the sex of the individual who carries the gene.

The two X chromosomes in the female are homologous and, as in the autosome pairs, have two genes at each locus. In the male, with one X and one Y, the sex chromosomes are not homologous (Fig. 2-13). Theoretically, there are three possible types of sex-linked transmission: (1) the genes are carried on the X chromosome exclusively, (2) the genes are borne only on the Y chromosome with no corresponding locus on the X chromosome, and (3) the genes occur at paired loci on both the X and the Y chromosomes. This last possibility has not been satisfactorily demonstrated in man and will not be considered at this time. Genes located on the Y chromosome (Y-linked) seem to be specifically responsible for the male phenotype and carry no known medically significant genes. Some minor abnormalities have been recorded; such a condition (hairy ear rims reported in some Indian families) is called *holandric inheritance*. The group of sex-linked disorders of outstanding importance are those carried on the X chromosome (X-linked).

There are a large number of traits (60 to 70) known to be carried on the X

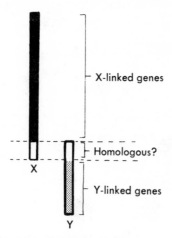

Fig. 2-13. The X and Y chromosomes.

chromosome. The outstanding characteristic of X-linked inheritance is that genes on the X chromosome have no counterpart on the Y chromosome. All females, with two X chromosomes, are homozygous or heterozygous for genes located on the X chromosome; but males, who have one X and one Y chromosome, have only one gene on the X chromosome (hemizygous). As a result, traits determined by the X-linked genes are *always* expressed in the male. For this reason the terms sex-linked and X-linked are usually synonymous. X-linkage is not to be confused with gene linkage, meaning two or more genes linked together on the same chromosome (Chapter 1). One of the most significant aspects of X-linked inheritance is the absence of father to son transmission. The father gives his one X chromosome to his daughters.

X-linked dominant inheritance

Few disorders exist today that are known to be inherited by the X-linked dominant mode of inheritance. One of these is vitamin D–resistant (hypophosphatemic) rickets, a skeletal disease in children that is due to a deficiency of vitamin D. The pedigree pattern (Fig. 2-14) superficially resembles the autosomal dominant disorders with a significant difference—an affected male transmits the trait to *all* of his daughters but to *none* of his sons. Half the sons and half the daughters of an affected female will display the abnormality (Fig. 2-15). Because there is no normal gene to modify the effect, males with the trait show symptoms of greater severity than females (the double-dose effect).

Briefly summarized, major characteristics of X-linked dominant inheritance are: (1) affected individuals will have an affected parent, (2) all the daughters but none of the sons of an affected male will be affected, (3) half the sons and half the daughters of an affected female will be affected, (4) normal children of an affected parent will have normal offspring, and (5) the inheritance pattern shows a positive family history.

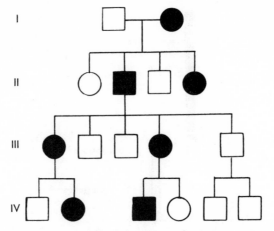

Fig. 2-14. An X-linked dominant inheritance pattern.

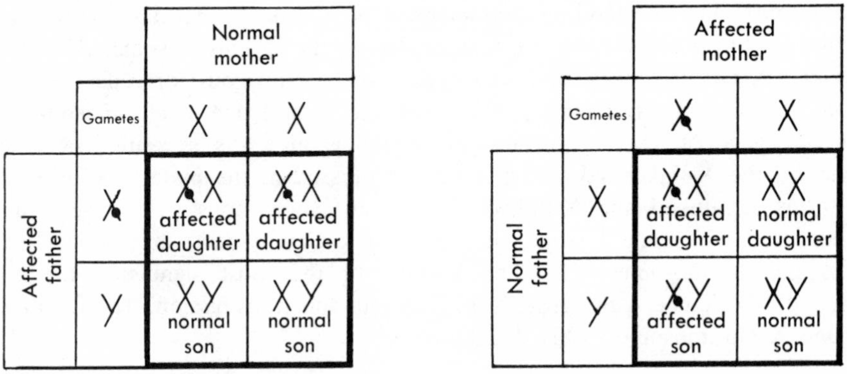

	Normal mother	
Gametes	X	X
Affected father ✗	X̶X affected daughter	X̶X affected daughter
y	XY normal son	XY normal son

	Affected mother	
Gametes	X̶	X
Normal father X	X̶X affected daughter	XX normal daughter
y	X̶Y affected son	XY normal son

Fig. 2-15. Punnett square illustrating the sex differences in offspring ratios in X-linked dominant inheritance. • = Dominant allele on X chromosome.

X-linked recessive inheritance

A number of serious conditions have been identified that are determined by a recessive gene on the X chromosome. The abnormal gene behaves as any recessive gene; that is, its effect will be opposed by a normal dominant allelic gene. The significant feature of this situation is that, with only one X chromosome, the male who receives an X-linked mutant gene will always display the disorder. He cannot transmit the disorder to his sons (Figs. 2-16 and 2-17). However, he does transmit the mutant gene to *all* of his daughters, who will be carriers of the gene. In contrast, a normal female, heterozygous for the mutant gene, will transmit the gene to half of her sons (affected) and half of her daughters (carriers).

The classic example of an X-linked recessive disorder is hemophilia, a disease due to an interference with the normal blood-clotting mechanism (p. 61). As a

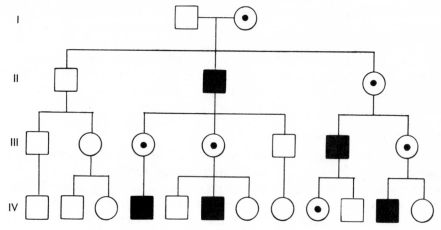

Fig. 2-16. An X-linked recessive inheritance pattern.

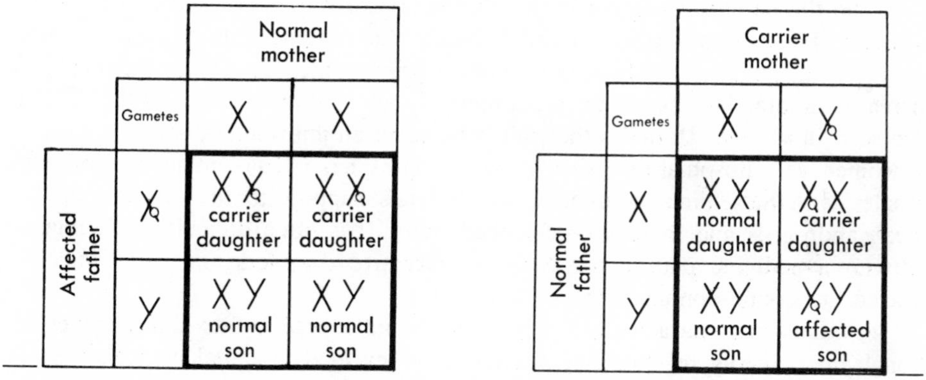

Fig. 2-17. Punnett square illustrating the sex differences in offspring ratios in X-linked recessive inheritance. o = Recessive allele on X chromosome.

result of prolonged clotting time the victims of hemophilia bleed excessively following injury or trauma and are commonly known as "bleeders." Queen Victoria of Great Britain transmitted this disorder to the Russian and Spanish royal families, with considerable political consequences. A few cases of hemophilia have been reported in women. Contrary to the usual expectation for a disorder in the homozygote, females with an X-linked disease are no more severely affected than males. This is due to *X inactivation,* discussed later in this section. An X-linked disorder in a female usually results from the mating between a carrier mother and an affected father. In such a situation all the sons of the affected female will be affected and all of her daughters will be carriers.

Other examples of disorders known to be transmitted as an X-linked recessive trait are the Duchenne-type muscular dystrophy, diabetes insipidus (both nephrogenic and pitressin-deficient), and one form of hydrocephalus.

Characteristics of an X-linked recessive disorder can be summarized: (1) affected individuals are principally males, (2) affected individuals will have unaffected parents, (3) half of the female siblings of an affected male will be carriers of the trait, (4) unaffected male siblings of an affected male cannot transmit the disorder, (5) sons of an affected male are unaffected, (6) daughters of an affected male are carriers, and (7) the unaffected male children of a carrier female do not transmit the disorder.

Sex-related gene expression

The sex of an individual may be a factor in the type and frequency of gene expression. In some conditions the genes are transmitted in the predictable manner according to autosomal inheritance patterns; however, their expression is limited to only one sex or the expression is different in the male than in the female.

Sex-influenced. An autosomal trait that is expressed more frequently in one sex is called *sex-influenced;* in the extreme, when only one sex is affected the gene is said to be *sex-limited* (rare). Sex-influenced traits are seen more frequently in the male and the influencing factor in these traits is probably the presence of testicular hormone. For example, gout, a painful disease due to deposition of purine crystals in the digital joints, is rarely seen in women prior to menopause but the incidence increases in later life; males are predominantly affected by baldness, although it is also seen in women. In males the trait behaves in an autosomal-dominant pattern, in women as autosomal-recessive. Also, baldness has been seen in heterozygous females with masculinizing tumors, which lends support to the relationship between androgens and these sex-influenced traits. It is often difficult to distinguish between a pedigree pattern for X-linked recessive disorders and these male-restricted autosomal-dominant traits.

Secondary sex characteristics appear to be influenced by sex-linked genes although undoubtedly regulated by sex hormone levels. The actual interrelationship seems unclear.

Chromosome inactivation

In X-linked disorders the homozygous female seems to be no more seriously affected than the hemizygous male although a more drastic effect would be expected in the female with a double dose of the abnormal gene. It has also been noted that there is no difference between normal males and normal females in the level of detectable enzymes that are products of X-linked gene action (for example, antihemophilic globulin). Again, it would be expected that genes on two X chromosomes would produce more than those on only one. The explanation for these observations is the *Lyon hypothesis* of X chromosome inactivation. Early in embryonic life (approximately the sixteenth day of gestation) one X chromosome in every somatic cell of the female becomes genetically inactivated (Fig. 2-18). This inactive X chromosome becomes very tightly coiled and can be observed in the interphase nucleus as a dark-stained *chromatin mass* or *Barr body.* It is normally seen only in the female somatic cells. Evidence gathered to date indicates that chance events alone determine whether the maternal or paternal X

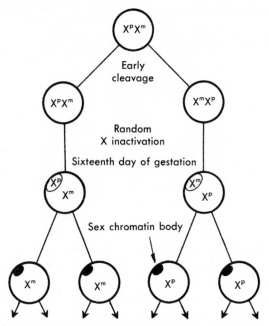

Fig. 2-18. Model of the Lyon hypothesis of X inactivation. The superscripts m and p indicate the maternal and paternal X chromosomes.

chromosome is inactivated in any given cell. Once inactivation of a particular X chromosome takes place, however, the same X chromosome remains inactive through all subsequent generations of that cell.

VARIATIONS IN INHERITANCE PATTERNS

There are a number of variables that may modify the basic inheritance patterns. The degree to which a gene exerts its effect or environmental factors may influence the phenotype.

Degrees of expression

Penetrance. In addition to the quantitative factor that indicates that genes in double dose tend to be more severe than in single dose, there is a qualitative factor that describes the strength of a gene. Often traits show a high degree of variability in expression from individual to individual, especially in autosomal-dominant disorders. When a gene produces its effect on the phenotype each time it is present in the genotype, it is said to be *fully penetrant* or to exhibit *complete penetrance*. It seems to be an all-or-none reaction. Achondroplasia, a form of dwarfism due to a defect in ossification, is always evident whenever the gene is present. If a condition is not recognized in the individual despite the fact that he carries the gene, it is said to be *nonpenetrant*. This accounts for what appears to be skipped generations seen in some pedigrees. Evidence of the gene can often be found, but the manifestation is reduced to such a mild degree as to be undetected except by very careful examination. Some disorders may have a partial or reduced penetrance. For example, retinoblastoma, a tumor of the retina due to a dominant gene, has 90% penetrance because it does not develop in 10% of the children with the gene.

Expressivity. On the other hand, *expressivity* refers to the degree of severity seen in the phenotype of certain individuals of a particular genotype. This variation in expression is especially pronounced in autosomal-dominant traits and seems to fall into a bell-shaped distribution curve with nonpenetrance at one end, complete penetrance at the other end, and varying degrees of expression in between. An example of expressivity is the autosomal-dominant trait osteogenesis imperfecta, a disorder characterized by blue sclera, bone fragility, and often otosclerosis. The manifestations of the disorder may be so mild in one person that it does not interfere with his activities yet so severe in another that he becomes totally incapacitated.

Varied expressivity often occurs with supernumerary digits. Polydactyly may appear on a hand or a foot and the digits may vary in number.

Anticipation. Another phenomenon associated with variability of expression in autosomal-dominant traits is *anticipation*. This means that with each successive generation the disorder manifests itself at an earlier age and with greater severity than in the previous generation. An example often cited is mytonic dystrophy, a progressive weakness and wasting of voluntary muscles accompanied by myotonia (delayed relaxation of muscles after contraction). In the first generation the effects may be relatively mild, the next generation will probably experience a wide range of severity, and their children will more likely display the disease earlier and be more severely affected than the generations before. Geneticists are not in agreement regarding anticipation and can offer no biologic basis for its existence.

Pleiotropy. Many genetic diseases are syndromes; that is, there may be a variety of phenotypic effects that are consistently associated with a particular disease. When a single gene produces multiple seemingly different and unrelated effects it is often referred to as *pleiotropic*. The varied clinical features observed in arachnodactyly, or Marfan's syndrome, are the result of a basic defect in the elastic fibers of connective tissue that causes such diverse effects as dislocation of the optic lens, aortic aneurysm, and skeletal deformities. The name arachnodactyly is derived from the characteristic appearance of those affected: tall, slender individuals with long, tapering fingers and toes. The musculature is poorly developed, and some of the skeletal deformities include kyphosis, scoliosis, pigeon chest, winged scapula, and arched palate. The number and severity of defects expressed varies in individuals.

Many autosomal-recessive disorders exhibit multiple and diverse effects. For example, the absence of the enzyme galactose-1-phosphate-uridyl transferase results in the aminoaciduria, cataracts, liver cirrhosis, and mental retardation of *galactosemia*. However, these are end results produced in normal tissues rather than direct gene action.

Phenocopy and genocopy

The phenotype is not necessarily an indication of the genotype. Environmental factors can produce effects that imitate or mimic those determined by a mutant gene. *Phenocopy* is the term used to describe a condition due to an environmental factor (viruses, chemicals, radiation) that may be indistinguishable from a genetic disorder. Many of these are found in man and include such defects as cataracts, deafness, and microcephaly. A common illustration of phenocopy is a darkly sun-tanned individual with normally lightly pigmented skin who has the same skin color as one with genetically dark pigmentation. Since differentiation between hereditary and environmental causes is of importance in prevention, diagnosis, and counseling it will be discussed in greater detail later.

Also, different genes may produce identical effects in the phenotype. The terms *genocopy* and *mimic genes* describe this possibility. One example is hereditary deafness; another is the muscular dystrophies. The genes for both of these disorders may produce similar effects but are inherited by any of the major inheritance patterns—autosomal-recessive, autosomal-dominant, and X-linked recessive. A distinction is essential, especially for purposes of genetic counseling.

REFERENCE

1. McKusick, V. A.: Mendelian inheritance in man, ed. 3, Baltimore, 1971, The Johns Hopkins University Press.

GENERAL REFERENCES

Crispins, C. G., Jr.: Essentials of medical genetics, New York, 1971, Harper & Row, Publishers.

Gardner, E. J.: Principles of genetics, ed. 4, New York, 1972, John Wiley & Sons, Inc.

McKusick, V. A.: Human genetics, ed. 3, Englewood Cliffs, N. J., 1969, Prentice-Hall, Inc.

Roberts, J. A. F.: An introduction to medical genetics, ed. 5, London, 1970, Oxford University Press, Inc.

Stern, C.: Principles of human genetics, ed. 3, San Francisco, 1972, W. H. Freeman and Co., Publishers.

Thompson, J. S., and Thompson, M. W.: Genetics in medicine, Philadelphia, 1966, W. B. Saunders Co.

3

Some single-gene disorders

There is an abundant variety of diseases, each of which is known to be due to a single mutant gene. They consist of those disorders for which the responsible mutant gene has been identified through observation of its distribution patterns in families, or kindreds, in which the disorder has been displayed. Single genes are readily recognized as hereditary inasmuch as they are distributed in kindreds according to the simple mendelian principles of inheritance, and they are frequently classified according to these inheritance patterns—dominant, recessive, and X-linked. A more satisfactory classification is one that identifies the basic chemical defect or the process in which its primary effect is manifest. Much has been learned about the nature of normal functions and their genetic control through the study of these abnormalities of nature. Most of these diseases, though rare, provide useful illustrations of the manner in which a gene acts to produce single or multiple phenotypic effects.

End products of gene action are primarily proteins, either structural cell components or enzymes. The molecular basis of the genotype is DNA, and the unit, or segment, of this genetic material known as the gene is responsible for the production of a specific protein (p. 16). These proteins (polypeptide chains, or enzymes) form the molecular basis of the phenotype. There appears to be a one-to-one relationship between genes and specific end products—when the gene is present the protein is present. Genes are potentially mutable units; therefore, any change in a gene will produce a disturbance in protein synthesis and lead to an alteration in the process or processes dependent upon it. A disturbance in the normal processes produces a phenotypic effect of greater or lesser consequence to the organism. This chapter is concerned with those phenotype alterations that are expressed as disease in that they cause disagreeable symptoms or undesirable physical abnormalities.

Most single-gene disorders can be broadly grouped according to the type of protein altered by defective gene action. *Molecular disease* is a term sometimes applied to a defect in a structural protein (such as the hemoglobin molecule), while enzyme defects are considered to be *inborn errors of metabolism*. All genetically determined disorders are essentially molecular diseases since a gene determines the molecular structure of any protein. At the same time, a gene that controls protein structure will frequently have some effect on metabolism. The selection and classification of the examples used in this chapter are not necessarily limited to a single category. In some disorders the basic defect may apply to one category, while the major clinical manifestation is related to another; for example, glucose-6-phosphate dehydrogenase deficiency is basically a disorder of glucose metabolism but is expressed primarily as a blood disorder.

INBORN ERRORS OF METABOLISM

At the turn of the century Sir Archibald E. Garrod demonstrated the relationship between genes and some specific metabolic reactions. By observing certain disorders of lifelong duration, he developed the concept that these diseases were due to lack of or reduced activity of an enzyme governing a specific step in a metabolic process. The familial incidence of these disorders led to the correct deduction that they were due to a hereditary lack of or deficiency in the responsible enzyme. Inborn errors of metabolism is a term used to describe familial diseases that are characterized by abnormal protein, carbohydrate, or fat metabolism.

Metabolism is the sum of all the physical and chemical processes involved in the growth, maintenance, and transformation of organized substance in living systems. All the biochemical processes are under genetic control and consist of a complex sequence of reactions. Each step in a metabolic pathway is catalyzed by an enzyme. When, through defective gene action, the specific required enzyme is absent or deficient the normal biochemical process is blocked, resulting in phenotypic effects of greater or lesser consequence. This relationship is the one gene–one enzyme concept of later geneticists Beadle and Tatum. Recently this principle has been modified to state the more accurate concept one gene–one polypeptide, thus including structural proteins as well as enzymes.

The mutation of a single gene produces an alteration only in that specific protein for which it is responsible. When this protein, an enzyme, controls one of a series of chemical reactions, it can also affect the quality or quantity of other reactions. Fig. 3-1, *A,* schematically represents a normal metabolic pathway; Fig. 3-1, *B,* illustrates how a change in hereditary material that interferes with the synthesis of an essential enzyme interrupts this process. A block in the normal pathway can produce an accumulation of products preceding the block, inhibit the function of enzymes in succeeding reactions, or create a deficiency in the product. Sometimes alternate pathways are used and the products of these processes increases. One or all of these effects of gene action are observable in the individual as disease. It has been suggested that all human beings suffer from one error of metabolism, that is, the inability to synthesize ascorbic acid (as other species do). The deficiency disease is scurvy and the remedy is supplementary vitamin C in the

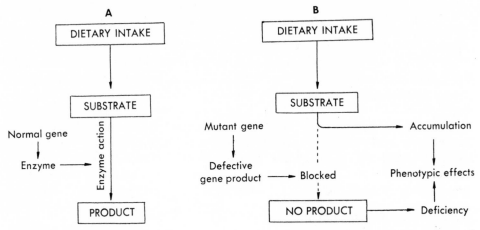

Fig. 3-1. A metabolic pathway. **A,** A normal metabolic pathway. **B,** The effect of defective gene action.

diet. In the same way, the adverse effects due to defects in metabolism have been modified by providing the deficient product to affected persons (for example, thyroid extract in cretinism). Also, some degree of success is achieved by decreasing the intake of the substance that the body is unable to properly metabolize, thus decreasing the accumulation of deleterious metabolites (for example, milk sugar in galactosemia).

There are many heritable disorders due to an inborn error of metabolism involving either the buildup or breakdown of metabolic processes. They are rare diseases, and the mode of inheritance is almost always autosomal-recessive. This is best understood by considering the double-dose effect as it relates to the gene-enzyme concept. If a gene is responsible for the formation of an essential enzyme and each individual has two such genes (the normal homozygote), then the enzyme is produced in normal amounts. The heterozygote, who has only one gene with a normal effect, will produce the enzyme in sufficient amounts to carry out the metabolic function under normal circumstances. Therefore, the heterozygote will not exhibit symptoms of the disorder, whereas the abnormal homozygote, who inherits a defective gene from both parents, will have no functioning enzyme and will be clinically affected.

DEFECTS IN AMINO ACID METABOLISM

The most frequently used model of a defect in amino acid metabolism involves the metabolism of the basic amino acid phenylalanine. Numerous enzymes are required for the complex processes that convert phenylalanine to a variety of biochemical products. The sites of metabolic blocks in several disorders are illustrated in Fig. 3-2, but for simplicity many intermediate steps in the process are omitted. Absence of an enzyme at any point in the pathway will in some way disturb the reactions dependent upon it. A deficient enzyme at point A prevents the conver-

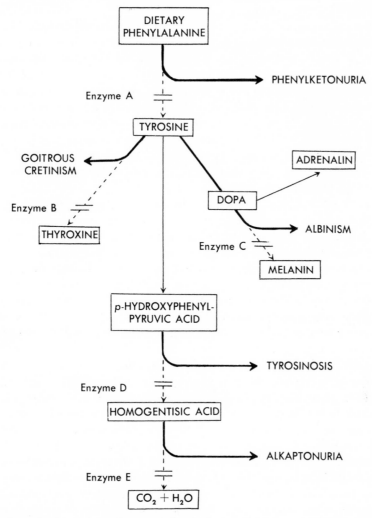

Fig. 3-2. Some basic steps in the metabolism of phenylalanine and the consequences of blocks at various steps in the metabolic pathway.

sion of phenylalanine to tyrosine and results in *phenylketonuria* (PKU). A defective enzyme at point B will produce a familial form of *goitrous cretinism*. If there is a genetic block in biochemical conversion at point C, melanin pigment is not formed and the individual will lack skin, hair, and eye color in varying degrees—the well known *albinism*. *Tyrosinosis*, a very rare disorder, is due to a block at point D, and *alkaptonuria* occurs when enzyme E prevents breakdown of homogentisic acid to carbon dioxide and water. Large amounts of homogentisic acid excreted in the urine cause it to turn black upon exposure to air. It is this relatively benign disorder that Garrod described and that provided the basis for his famous concept of inborn errors of metabolism.

Phenylketonuria

PKU serves as an excellent example of an inborn error of metabolism. The enzyme defect has been determined, the mode of inheritance has been established, a form of therapy is available, heterozygotes can be identified, and widespread screening procedures are carried out for the early detection of affected infants.

HEREDITY AND INCIDENCE. Phenylketonuria is inherited as an autosomal-recessive disorder; therefore, the affected individual receives an abnormal gene from each of his heterozygous parents. The incidence in the population is estimated to be 1 in 10,000 to 20,000 births and accounts for approximately 1% of institutionalized mentally deficient persons. It affects primarily Caucasians, with the incidence being highest in Northern Europe and the United States, and is very rare in African, Jewish, and Japanese populations.

CHARACTERISTICS. Affected children appear normal at birth, and almost all have fair skin, blonde hair, and blue eyes. Although the age of appearance varies, untreated infants with PKU fail to thrive, and eczematous rashes, vomiting, and irritability are common. The most important clinical characteristic of PKU is severe mental retardation, but it is uncertain exactly when retardation begins. It is usually unrecognized from clinical symptoms until 6 months of age but rapidly increases in severity during the first year of life. Many of these children have seizures (especially the more severely retarded), are usually hyperactive, and exhibit erratic and unpredictable behavior. The majority have abnormal electroencephalograms.

BASIC DEFECT. When the liver enzyme *phenylalanine hydroxylase* is deficient, phenylalanine is not converted to tyrosine. As a result, phenylalanine and other metabolites of phenylalanine (formed through alternate pathways) reach high concentrations in the blood, urine, sweat, cerebrospinal fluid, and tissues. (Actually, these are not abnormal metabolites but normal ones in abnormal amounts). Heterozygotes, with only one defective gene, can produce enough enzyme to prevent the deleterious effects although the abnormal metabolites can be detected in their blood. The homozygote has only abnormal genes and is deficient in enzyme production. In an infant receiving a normal diet the blood phenylalanine levels will increase from the normal 1 to 4 mg/100 ml to levels of 10 to 60 mg/100 ml within the first week of life. When the serum phenylalanine level becomes sufficiently elevated a metabolite, *phenylpyruvic acid,* can be detected in the urine. The urine has a musty odor, and a few drops of 5% ferric chloride placed on a wet diaper will immediately turn a green color (the more convenient Phenistix test tapes can also be used). This test is not reliable until the infant is approximately 4 to 6 weeks of age, however.

Most of the clinical features in PKU can be attributed to the excessive accumulation of these products. For example, the decreased tyrosine leads to a diminished melanin production and the reduced pigment in skin, hair, and eyes. The specific mechanism that produces mental retardation is unknown, but it is probably due to a neurotoxic agent that has an inhibitory effect on development before myelinization in the central nervous system is complete, and damage to the brain may have already begun by the time the phenylpyruvic acid is detected in the urine. There are, however, instances in which individuals with elevated phenylalanine levels are of normal intellect.

DIAGNOSIS. The occurrence of the disease is suspected on the basis of finding phenylpyruvic acid in the urine and confirmed by detecting elevated serum phenylalanine levels. This simple blood test can be performed before the infant leaves the hospital, and many states require it as a routine screening procedure. Normal-appearing heterozygous carriers can often be identified by means of a phenylalanine tolerance test. Most heterozygotes have higher plasma phenylalanine levels than do normal persons.

TREATMENT. The genetic enzyme defect is intracellular; therefore, administration of the missing enzyme systemically is of no value. Instead, treatment is aimed at eliminating from the diet those foods that contain high concentrations of phenylalanine. It is felt by those who advocate the dietary regime that when this phenylalanine-restricted diet is begun in early infancy (before 3 months of age) and maintained during the period of most rapid myelinization, damage to the central nervous system can be averted. Virtually all animal and vegetable protein contains phenylalanine. In infancy, milk (which contains large amounts) is replaced by a special protein formula* containing very little phenylalanine but all other essential amino acids. Supplementary foods, calculated for phenylalanine equivalents, are gradually added to the diet and maintained at sufficiently low levels that the serum phenylalanine remains within safe limits. Most experts feel the child should remain on the diet during the period when the brain is most susceptible to damage, so it is usually continued until at least 3 to 4 years of age.

Initiating the diet after brain damage has occurred does not reverse the process but, if begun early enough, may limit its progress. Restricting phenylalanine in older children with PKU has indicated improvement in symptoms such as eczema and some behavior problems. A relatively recent finding concerns mothers with PKU who have given birth to mentally retarded children who do not have PKU. It is suggested that damage to the fetal brain results from transplacental transfer of maternal phenylalanine; therefore, these mothers should probably be placed on low-phenylalanine diets during pregnancy.

Albinism

The biochemical defect in albinism appears to be absence of the enzyme *tyrosinase,* which prevents the synthesis of melanin by the pigment-forming cells (melanocytes). These individuals have very white skin, fine white hair, and pink or light blue irises. Associated with these readily recognizable characteristics are numerous ocular abnormalities, including strabismus, refractive errors, nystagmus, photophobia, and poor visual acuity.

One of the more common of the rare metabolic disorders, the disease is inherited as an autosomal-recessive trait and the overall incidence is approximately 1 in every 5,000 to 1 in every 20,000 people. This incidence shows wide variability and is very high in some racial isolates. It is seen in all races and is especially

---

*Lofenalac, Mead Johnson Laboratories, Evansville, Indiana.

striking in the dark-skinned races. There is no treatment for albinism, and therapy is directed toward alleviating the ocular disturbances.

Goitrous cretinism

Hypothyroidism is one of the most common abnormalities of childhood, but only a small portion of cases are due to hereditary block in amino acid metabolism. The biochemical defect is absence of the enzyme *iodotyrosine deiodinase* and, therefore, failure to convert inorganic iodine into organic iodine. The symptoms are those seen in hypothyroidism of any etiology: stunted growth, lethargy, coarse hair, poor muscle tone, and typical facial features such as flat upturned nose, large, thick protruding tongue, and open mouth. Mental and physical development is delayed, and all patients will eventually develop goiter. The inheritance pattern of goitrous cretinism is autosomal-recessive as are the other heritable forms of cretinism that are due to other defects related to deficient thyroid production. The treatment is administration of thyroid extract.

DEFECTS IN CARBOHYDRATE METABOLISM
Galactosemia

Galactosemia is an inborn error of carbohydrate metabolism in which the body is unable to utilize the sugars galactose and lactose. Normally, lactose (milk sugar) is broken down in the gastrointestinal tract into galactose and then, following absorption, is enzymatically converted to glucose in the liver. When one of the enzymes responsible for this process, *galactose-1-phosphate uridyl transferase,* is absent, galactose accumulates in the blood and tissues.

CHARACTERISTICS. The disease is inherited as an autosomal-recessive trait and the incidence in the population is approximately 1 in 25,000 to 35,000 although with earlier recognition it may be higher. Infants with this disorder appear normal at birth but within a few days begin to vomit and lose weight. Jaundice and liver enlargement are often early signs reflecting the liver involvement that may progress to cirrhosis. Other symptoms are drowsiness, nausea, and diarrhea. Death often occurs in severe cases, and the untreated infants who survive develop irreversible cataracts and mental retardation (Fig. 3-3).

DIAGNOSIS. Diagnosis is made on the basis of galactosuria, increased levels of galactose in the blood, or by measuring the levels of galactose-1-phosphate uridyl transferase in red blood cells. Heterozygotes can be recognized by the decreased level of the enzyme in these erythrocytes.

TREATMENT. Treatment consists of eliminating galactose from the diet, that is, all milk and galactose-containing foods. This also includes many pharmaceutical products that contain lactose as fillers.

Glucose-6-phosphate dehydrogenase deficiency

Hereditary deficiency of the red blood cell enzyme *glucose-6-phosphate dehydrogenase* (G-6-PD) is relatively harmless under normal conditions. However, affected individuals develop acute hemolytic anemia under certain circumstances: following administration of primaquine (an antimalarial drug); ingestion of the

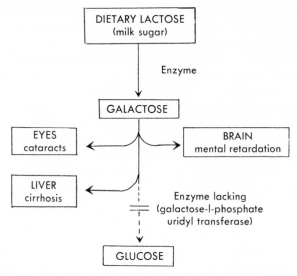

Fig. 3-3. The development of galactosemia.

fava bean (favism); or exposure to the compounds acetanilid, the sulfonamides, nitrofurantoin, or naphthalene (mothball poisoning). There is wide variability in the extent of sensitivity to the hemolytic action of these and other drugs. Manifestations are hemoglobinuria, hemolytic anemia, and jaundice for several days following exposure to these compounds. Recovery is usually spontaneous within a few weeks and death is uncommon.

The disorder is inherited as an X-linked recessive trait and is relatively common in some populations. The varieties seen most frequently in Western populations affect Afro-Americans and Caucasians from the Mediterranean regions. The form affecting black males is less severe than the Mediterranean form but it is estimated that probably 11% to 14% of Afro-American men are hemizygous for the trait.

Glycogen storage diseases

The deposition of glycogen in tissues characterizes a group of disorders collectively known as glycogen storage diseases. They are all genetically and biochemically distinct disorders although many are clinically similar. All are relatively uncommon autosomal-recessive traits and are classified according to the organs involved or the deficient enzyme. All interfere with conversion of glycogen to glucose in the affected tissues.

Lack of the enzyme *glucose-6-phosphatase* can cause *von Gierke's disease* (Type I), and organs affected are the liver, kidney, and gastrointestinal tract. Clinical manifestations include enlarged liver, hypoglycemia, and acidosis from formation of lactic acid in the tissues. The prognosis is poor but manifestations become milder if the affected infant survives the first year of life.

Pompe's disease (Type II) is due to a deficiency of the enzyme *acid maltase*, with a generalized tissue involvement. Primary clinical manifestations are enlarged heart, heart failure, general muscle weakness, and death in infancy.

In *McArdle's disease* (Type IV) there is absence of *muscle phosphorylase,* resulting in skeletal muscle pain and stiffness. There is fatigue in childhood and adolescence, weakness and severe cramps on exertion in adulthood, and wasting and weakness in middle age.

DEFECTS IN LIPID METABOLISM
Tay-Sachs disease

Sometimes known as *amaurotic familial idiocy,* this is a disease of infancy characterized by accumulation of the lipid *ganglioside* in all the neurons of the central nervous system and results in arrested development, spasticity, gradual loss of vision, and dementia. A characteristic cherry red macular spot can be observed by ophthalmic examination and confirms the diagnosis. Most affected infants appear normal for the first months of life. At approximately 4 to 6 months of age they develop apathy and general regression in motor and social development, then progressively deteriorate until death at about 3 to 4 years of age.

This disorder is inherited as an autosomal-recessive trait and primarily affects Jewish families, particularly those of Northeastern European origin. In affected individuals there is a deficiency of the enzyme *hexosaminidase A* although its precise effect is unclear. Some success has been achieved in detecting heterozygotes, who demonstrate a lower level of this enzyme, and in diagnosing affected fetuses in utero.

Juvenile amaurotic familial idiocy

Also transmitted as an autosomal-recessive trait, this disorder has characteristics similar to the infantile variety but is distinguished by later onset of initial symptoms (not earlier than 3 years of age) and mental and physical degeneration more often proceeds at a slower rate. Vision is usually gone within 2 to 4 years but the central nervous system symptoms (for example, seizures, disorders of gait, dementia, and wasting) progress until death. Unlike the infantile form, this disease is uncommon in Jews.

DEFECTS IN RENAL TRANSPORT MECHANISMS
Nephrogenic diabetes insipidus

In this disorder there is a failure of the renal tubules to respond to antidiuretic hormone and thus to reabsorb water. There is a continual excretion of urine, accompanied by excessive thirst. If there is insufficient water intake the individual rapidly becomes dehydrated and feverish. This is particularly hazardous in infants because thirst is difficult to assess and fluid losses are greater in relation to total body mass. Death sometimes occurs before the condition is recognized. All other renal functions are normal, and when hydration is adequate, the primary symptoms are excessive thirst (polydipsia) and excretion of large volumes of urine (polyuria) with abnormally low specific gravity.

The mode of inheritance is somewhat unclear. It appears to be X-linked since there is a higher incidence in males; however, many females also have the disorder. There are suggestions that it may be an autosomal-dominant trait with decreased penetrance in the female. The heterozygous female definitely seems to reflect a partial defect. Female relatives of affected males have shown decreased ability to concentrate urine, which may identify heterozygous carriers. The incidence is probably 1 in 10,000 to 20,000 births.

Therapy is directed toward prevention of dehydration and its complications. Sometimes drinking large volumes of water interferes with food intake, which may be reflected in subnormal growth. Paradoxically, the thiozide diuretics have proved to be of value in reducing the polyuria, although the mechanism involved is not clearly understood.

Renal glycosuria

This is a relatively benign disorder in which there is a defect in the renal transport mechanism for glucose and, therefore a failure to reabsorb glucose in the proximal tubules. It is characterized by excretion of glucose in the presence of normal glucose level and is inherited as an autosomal-dominant trait. No treatment is indicated, but the condition is important in differential diagnosis of the more serious disorder diabetes mellitus.

Vitamin D—resistant rickets

This disorder is one of the rare conditions due to an X-linked dominant gene in which there is a decreased reabsorption of inorganic phosphate from the renal tubules. The result is hypophosphatemia and decreased absorption of calcium from the intestine, causing the skeletal deformation of rickets in small children and osteomalacia in adults. Massive doses of vitamin D seem to be an effective therapy.

DEFECTS IN PORPHYRIN METABOLISM

Porphyrin is an essential component of numerous substances concerned with internal respiration. For example, iron porphyrin composes the heme portion of hemoglobin and serves as the oxygen transport mechanism in blood. In one group of heritable disorders the body fails to metabolize porphyrins and their precursors properly, and thus there is an increased excretion of these compounds and their by-products. Two general classifications are identified: the congenital, or erythropoietic porphyria, in which the metabolic defect occurs in the bone marrow, and the hepatic porphyrias due to a defect in liver metabolism.

Hepatic porphyrias

There are at least three hereditary varieties of hepatic porphyria. The incidence of *Porphyria variegata* is especially high in the white South African population. It and two other types, *intermittent acute porphyria* (Swedish type) and *coproporphyria,* are all transmitted by an autosomal-dominant inheritance pattern. The basic defect seems to be an increased activity of the liver enzyme *ALA (6-aminolevulinic acid) synthetase,* and all have common features: initial appearance of clinical manifestations occurs during late puberty or early adulthood, exacerbations of the symptoms may be precipitated by certain drugs in therapeutic doses, and all are associated with the same neurologic syndrome.

The presence of increased porphyrins in stool and urine, easily detected with ultraviolet light, provides a simple screening device. During an attack the urine becomes the color of port wine. The most important aspect of treatment of the porphyrias is prevention of attacks.

Acute intermittent porphyria. Under ordinary circumstances no more serious effects are encountered in this condition than vague complaints such as dyspepsia or nervousness. However, after taking barbiturates or some sulfonamides, or after thiopentone anesthesia these individuals develop acute abdominal and muscle pain, vomiting, rapid pulse, emotional disturbances, and peripheral neuritis that often leads to total paralysis and death. In an acute attack the person's life is in peril. Some females with the disorder may complain of abdominal pain in the premenstrual period or during pregnancy, frequently resulting in abdominal surgery where they are most apt to be given a barbiturate. Most persons recover from an attack, but the rate of recovery is to a great extent dependent upon the degree of peripheral neuritis and paralysis. It is this disorder that is determined to have been responsible for King George III's episodes of "madness."[1]

Porphyria variegata. This milder form of porphyria is characterized by a family history of chronic skin involvement. There is increased sensitivity of the skin to minor trauma and exposure to sunlight (photosensitivity). Acute abdominal and neurologic manifestations similar to those in acute intermittent porphyria may also be present at times, usually precipitated by ingestion of drugs, particularly the barbiturates.

Persons with *coproporphyria* are asymptomatic, or the symptoms may be similar to either of the previous hepatic varieties, although skin involvement is uncommon. Barbiturates cause acute attacks that, unlike the others, may occur at any age.

Congenital erythropoietic porphyria

This severe rare childhood form occurs less frequently than the hepatic porphyrias and has a wide racial distribution. Photosensitivity is usually present after the child is exposed to sunlight, with wide variation in degrees of skin involvement. Porphyrins can be detected in the urine, and an enlarged spleen is a constant feature.

DEFECTS OF STEROID METABOLISM
Adrenogenital syndromes

These disorders are characterized by overproduction of androgens by the adrenal cortex. Cortisol synthesis requires numerous chemical reactions, each step mediated by enzyme action. The adrenogenital syndromes are produced by deficiency of one of these enzymes. A decrease in cortisol stimulates the production of corticotrophin (ACTH) by the pituitary gland. The adrenal glands respond to ACTH stimulation by producing all other hormones for which they are responsible, especially the androgens and aldosterone. Phenotypic effects are the result of increased secretion of these hormones.

Female infants with the disorder show masculinization, with enlarged clitoris and variable degrees of labial fusion. This syndrome should be considered in any situation where sex is doubtful. If untreated, the affected females progressively develop other masculine features such as male hair distribution, deep voice, and further enlargement of the clitoris. Males show no abnormalities at birth but will show signs of excessive body development similar to precocious puberty at 4 to 5 years of age. Male testes and female sex organs are unaffected or may even fail to develop. The inheritance pattern is autosomal-recessive and the treatment is administration of cortisone.

SOME OTHER DISORDERS OF METABOLISM
Mucopolysaccharidoses, or lipochondrodystrophies

Two very similar disorders that seem to be associated with abnormalities of mucopolysaccharides are *Hurler's* and *Hunter's syndromes,* sometimes called *gargoylism* or *multiple dystrophies.* Most affected individuals are mentally defective. Their appearance is characterized by the distinctive facial appearance of broad head with widely spaced eyes, thick lips, and protruding tongue. There are skeletal deformities, including dwarfism and curvature of the spine plus an enlarged liver and spleen. The inheritance patterns are of two types, a more severe autosomal-recessive (Hurler's syndrome) and an X-linked recessive (Hunter's syndrome). For counseling purposes, the difference is reflected in the observable corneal clouding seen in the autosomal-recessive form. Most afflicted people do not survive childhood.

Cystic fibrosis of the pancreas

Cystic fibrosis, fibrocystic disease of the pancreas, and *mucoviscidosis* are all terms applied to this important and most common heritable disease of children, adolescents, and young adults. It is one of the most serious chronic diseases of Caucasian children and accounts for most nontuberculous lung disease in this group. In addition to the chronic pulmonary disease there is severe malabsorption due to pancreatic deficiency and abnormally high sweat electrolyte levels. Although there is some variability in manifestations, the morbidity is severe and the mortality high. Most affected individuals do not survive early childhood, although more of the less severe cases are being recognized in the older age groups. A few affected females have borne offspring, but affected males are almost always sterile.

HEREDITY AND INCIDENCE. Cystic fibrosis is inherited as an autosomal-recessive trait; therefore, the victims are homozygotes who received the mutant genes from heterozygous parents. The incidence of the fully manifested disease is estimated at 1 in every 1,500 to 2,000 births in predominantly Caucasian populations. The disease is almost nonexistent in the Mongoloid races, and the rare affected blacks are usually in areas where there is apt to be mixed ancestry. The carrier rate for the recessive gene is computed at 1 in 20 to 25 persons. Because most affected individuals never reproduce, this high frequency must be maintained by either a high mutation rate or a selective advantage in the heterozygote. Some studies have indicated that perhaps there is increased fertility related to the carrier state.

BASIC DEFECT. The basic biochemical defect in cystic fibrosis is unknown. It is assumed, because of the single-gene etiology, that it is probably due to alteration of a protein, perhaps an enzyme. It has also been proposed that the defect might be related to an abnormal function of the autonomic nervous system and hence overstimulation of the cholinergic glands of the body. So far there is no conclusive evidence to support any of these theories. The underlying mechanical defect is well known, however, and the factor responsible for the multiple clinical manifestations of the disease is the increased viscosity of mucous gland secretions. Instead of forming a thin, freely flowing secretion the mucous glands produce a thick, inspissated mucoprotein that accumulates and dilates them. Small passages in organs such as the pancreas and bronchioles are obstructed with precipitated or coagulated secretions.

In the pancreas the thick secretions block the ducts, leading to cystic dilations of the acini (small lobes of the gland), which then undergo degeneration and progressive, diffuse fibrosis. Because essential pancreatic enzymes are unable to reach the duodenum, digestion and absorption of nutrients are markedly impaired—particularly fats, proteins, and to a lesser degree carbohydrates. The islands of Langerhans remain unaffected but may be decreased in number with increased pancreatic fibrosis.

A unique characteristic of cystic fibrosis is the consistent finding of abnormally high sodium and chloride concentrations in the sweat. Mothers frequently observe that their infants taste "salty" when they kiss them. This increased electrolyte content is present at birth and persists throughout life and is unrelated to the severity and clinical course of the disease. The volume of sweat is not affected.

CHARACTERISTICS. Clinical evidence of cystic fibrosis varies from one individual to another, and the manifestations may range from relatively little pulmonary and digestive impairment to very severe lung and gastrointestinal disability. Usually both systems are involved, but the disease may be limited to difficulties with only one or the other. All manifestations of the disease are a direct or indirect result of obstruction of mucous ducts (pancreatic or bronchial) by thick mucous secretions.

In the *gastrointestinal tract,* absence of the pancreatic enzymes trypsin, amylase, and lipase prevent the conversion of ingested food into compounds that can be absorbed by intestinal mucosa. Consequently the nondigested food is excreted in excessively large, frothy, and extremely foul-smelling stools. Because so little is absorbed from the intestine the child compensates with a voracious appetite. The abdomen is very enlarged and lack of absorption causes a marked wasting of tissues with loss of weight and failure to grow. The extremities are thin, and the skin droops from wasted buttocks. Anemia is a common complication. The first indication of cystic fibrosis, seen during the newborn period in approximately 10% of infants, is *meconium ileus.* The abnormally thick, puttylike meconium stool obstructs the small bowel, requiring surgical intervention.

Pulmonary complications are present in almost all individuals affected with cystic fibrosis and constitute the most serious threat to life. Bronchial and bronchiolar obstruction by the thick mucus produces varying degrees of ventilatory dysfunction. These patients have a severe hacking cough, dyspnea, and the barrel

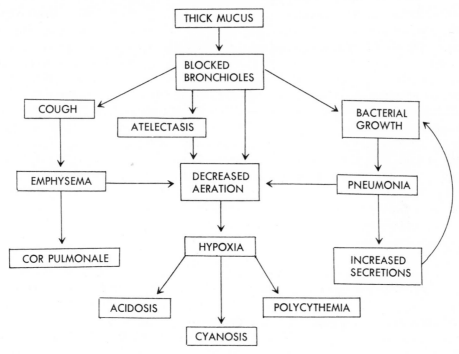

Fig. 3-4. Pulmonary effects of cystic fibrosis.

chest characteristic of chronic obstructive pulmonary disease. Atelectasis is common. The lung involvement leading to chronic hypoxia, hypercapnea, and acidosis frequently results in pulmonary hypertension and heart disease (cor pulmonale). The trapped mucus provides an excellent medium for bacterial growth contributing to frequent severe upper respiratory tract infections (Fig. 3-4).

DIAGNOSIS. Diagnosis is made on the basis of x-ray examination of the lungs, which indicates generalized emphysema, absence of trypsin and other pancreatic enzymes in the stool or duodenal contents, and evidence of increased sweat electrolyte concentrations. Testing for the latter is simple and reliable. Local stimulation of sweat on a small patch of skin (usually the forearm or leg) is accomplished by *iontophoresis* of pilocarpine. A small, painless electric current is used to carry the cholinergic drug into the skin to stimulate production of sweat, which is collected on a filter paper and the chloride content measured in the laboratory. A sweat chloride content over 60 mEq/100 ml is considered to be diagnostic of cystic fibrosis and between 40 and 60 mEq/100 ml is highly suggestive. Some investigators have observed that a number of relatives of patients have elevated sweat chloride levels, although not as high as affected individuals. (It is also suggested that there may be some relationship between heterozygosity and susceptibility to some upper respiratory tract disorders.)

TREATMENT. The therapy for gastrointestinal involvement is replacement of pancreatic enzymes. Since the enzyme deficiency is extracellular, replacement is achieved by oral administration together with a high protein–high carbohydrate diet and water-miscible vitamin supplements. Extra salt may be necessary during warm atmospheric temperatures and febrile states to replace that lost by excess sweating.

The pulmonary involvement requires vigorous efforts aimed at prevention and control of complications. Affected individuals are protected from exposure to upper respiratory tract infections whenever possible. Intensive antimicrobial therapy is instituted for control of infection, and some authorities employ it as a prophylactic barrier either continuously or during seasons of greatest threat. Most agree that routine high humidity therapy to facilitate the liquifying of bronchial secretions plus physical therapy to promote bronchial drainage are of value.

Progress is being made in methods to identify heterozygotes, and it is hoped that widespread screening procedures can be instituted to detect carriers, as is now possible with some other lethal and semilethal disorders.

DISORDERS OF THE BLOOD

This segment will be concerned only with diseases manifesting serious blood disorders that are transmitted by single-gene inheritance. There are other serious disorders related to normal blood under special circumstances (for example, erythroblastosis fetalis) that will be elaborated upon later. These single-gene disorders of blood involve defects in coagulation and the hemoglobinopathies.

DISORDERS OF COAGULATION

There is a hereditary deficiency involving each of the factors that contribute to the coagulation of blood. In order to understand the way in which a defect in the process produces its effect, it is necessary to have a basic knowledge of the normal clotting mechanism. Normal blood plasma contains a number of factors that, when triggered by trauma, progress through a chain of reactions ultimately to transform a soluble protein (fibrinogen) into an insoluble protein (fibrin)—the substance of the blood clot. One product serves as a catalyst, or stimulant, for the next reaction. If a factor is deficient or defective, the process is suspended at that particular phase and can proceed no further.

Although extravascular factors are important to blood coagulation (for example, the blood vessels and surrounding tissues), it is the intravascular factors that are very briefly outlined in Fig. 3-5. In *phase I,* thromboplastin precursors circulating in the plasma react with calcium and a substance released from the platelets to produce plasma thromboplastin. In *phase II,* thromboplastin becomes activated through stimulation of additional factors (V, VII, and X) and, again in the presence of calcium, this thromboplastin acts as a catalyst in the conversion of prothrombin to thrombin. In *phase III,* the plasma fibrinogen is transformed to fibrin through the action of thrombin in conjunction with a fibrin-stimulating factor. In all three phases of the clotting process the protein components interact in pairs, the product of one reaction serving as an enzyme to catalyze another.

Inherited deficiencies have been described for most of the factors essential to the normal clotting of blood. The most common and serious of the coagulation

Phase I   Phase II   Phase III

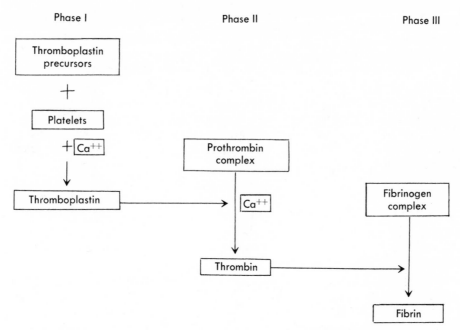

Fig. 3-5. The blood-clotting mechanism. See text for specific factors.

defects involve deficiencies of phase I factors and are frequently lumped under the large general category called the hemophilias. Phase I factors (plasma thromboplastin precursors) include antihemophilic globulin (AHG, or factor VIII), plasma thromboplastin component (PTC, or factor IX), plasma thromboplastin antecedent (PTA, or factor XI), and the Hageman factor (factor XII). Phase II factors (prothrombin complex) include labile factor (proaccelerin, Ac-globulin, or factor V), stable factor (proconvertin, or factor VII), Stuart-Prower factor (factor X), and prothrombin (factor II). Phase III factors include fibrinogen (factor I) and fibrin-stabilizing factor (factor XIII). Even the rarest of the bleeding disorders have been of great value to investigators in helping to sort out the complex components of the normal blood-clotting mechanism.

Hemophilia A

Classic hemophilia is one of the oldest hereditary diseases known to man. For centuries writers have been describing bleeders. Reference in the Talmud indicates that the ancient Jews recognized its hereditary nature and prohibited circumcision of infants whose family history suggested the disorder. This often uncontrollable bleeding is the most outstanding characteristic associated with this crippling disorder that affects males almost exclusively.

HEREDITY AND INCIDENCE. Hemophilia affects approximately 1 boy in 10,000, and the evidence for X-linked inheritance is firmly established. In all pedigrees studied, no instance of male-to-male transmission has been recorded. There is a positive family history in 80% of the cases; the remainder are probably the result of sporadic fresh mutations. Occasionally hemophilia occurs in the female. There are various means by which a female might be affected: mating of a female heterozygote to an affected male, fresh mutation in a male married to a heterozygous female, fresh mutation in a female married to an affected male, or fresh mutation in both parents. A female with an XO genotype could acquire the gene from either parent; however, most female bleeders are probably heterozygotes. These women have decreased plasma levels of antihemophilic globulin as compared with normal women, and the range of deficiency varies from near normal to severe, probably dependent upon the tissues in which the gene bearing X chromosome remains functional (see X inactivation, p. 42).

BASIC DEFECT. The basic defect of hemophilia A is a deficiency of the antihemophilic globulin (AHG), or factor VIII, necessary for the formation of plasma thromboplastin, and the severity of the disorder is directly related to the amount of AHG circulating in the plasma. When this factor is deficient or lacking, the coagulation is arrested at phase I. The precise source of AHG is still unknown.

CHARACTERISTICS. The outstanding manifestation of this disorder is uncontrollable bleeding from even minor trauma. It is evidenced early by a marked tendency to bruise. Hematomas develop in any tissue, viscera, or body cavity; and one of the most characteristic manifestations of hemophilia is bleeding into the joints, or hemarthrosis. The joints most frequently involved are the elbows, knees, and ankles, which become discolored, swollen, and painful. These hemorrhages may follow trauma or appear spontaneously, and repeated injury may produce degenerative changes and ultimately a fixed, unusable joint. Central nervous system and neck hemorrhages constitute life-threatening emergencies.

DIAGNOSIS AND TREATMENT. Diagnosis is made on the basis of family history associated with prolonged clotting time and low blood protein levels. Treatment consists of modifying or restricting activities to prevent trauma in affected individuals insofar as practical without imposing additional psychologic stresses. Blood transfusions have long been used as replacement therapy for blood loss and are still vital in treating hemorrhage. The prophylactic use of plasma concentrates and cryoprecipitates is valuable in prevention, although supplies are not yet adequate for effective widespread use.

Hemophilia B

Deficiency of factor IX—*Christmas disease,* or hemophilia B—is a genetically determined coagulation disorder clinically indistinguishable from hemophilia A. It also is inherited as an X-linked recessive trait but is a much rarer and milder disease than hemophilia A. Differential diagnosis is made on the basis of laboratory tests, and the treatment is administration of blood and plasma.

Other hemophilias

A rare disorder with a questionable inheritance pattern, which involves a capillary defect as well as a deficiency in at least one plasma protein, is *von Willebrand's disease,* or *vascular hemophilia.* It is probably an autosomal-dominant trait but seems to be more common in women than in men. The disease is characterized by nosebleeds, bruising, and bleeding following trauma, although the severity of the symptoms is highly variable even among members of the same family. There is a prolonged bleeding time but normal coagulation time.

Other rare inherited deficiency states appear to be mostly autosomal-recessive, including *congenital afibrinogenemia,* a phase III disorder characterized by a marked deficiency of fibrinogen. It is usually apparent at birth, and bleeding is the primary manifestation. Treatment consists of administration of concentrated fibrinogen and, when necessary, plasma or blood.

HEMOGLOBINOPATHIES

The major portion of the protein in red blood cells consists of hemoglobin, the element that carries oxygen and carbon dioxide. Increasing interest has been developing in regard to the structure and synthesis of this compound.

Normal hemoglobin is composed of a *heme,* or iron-carrying portion, bound to each of four polypeptides in the *globin* portion. These four chains occur in pairs with identical amino acid composition. In most adults the hemoglobin consists of two alpha (141 amino acid) chains and two beta (146 amino acid) chains and is known as hemoglobin A (HbA). As in all proteins, the ability or inability to synthesize normal hemoglobin is inherited from the parents. If the amino acid sequence (governed by gene action) is altered, the structure of the hemoglobin is also changed, which ultimately disturbs the physiologic function of these chains. The hemoglobinopathies are disorders due to genetically determined alterations in the molecular structure of hemoglobin.

Sickle cell anemia

A number of abnormal hemoglobins have been identified; the best known of these is sickle cell anemia. This severe, sometimes fatal disease derives its name from the bizarre shapes assumed by erythrocytes under conditions of low plasma oxygen concentration (Fig. 3-6).

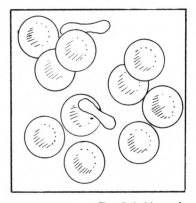

Fig. 3-6. Normal and sickled erythrocytes.

HEREDITY AND INCIDENCE. This disease occurs primarily in black populations, affecting approximately 1 in 500 to 600 black-Americans. The affected individuals are homozygous for the mutant gene, and the gene is paritally manifest in the heterozygote; therefore, the inheritance pattern is usually called intermediate.

BASIC DEFECT. In persons homozygous for the mutant sickle cell gene there is a substitution of one amino acid, valine, for the normal glutamic acid in the beta hemoglobin chain to form hemoglobin S (HbS) instead of HbA. This insoluble hemoglobin crystalizes under conditions of decreased oxygen concentration, causing the red blood cells to assume a crescent or sickle shape. These rigid cells accumulate and obstruct the flow of blood through small vessels to produce the tissue damage and painful symptoms of the sickle cell crisis. The decreased blood flow causes further oxygen depletion and more sickling to create a vicious circle. In addition, the body responds by rapidly destroying these fragile sickled cells before the process can be reversed, producing a severe anemia.

In the heterozygote, who possesses both the abnormal HbS and the normal HbA, the amount of HbS in individual red blood cells varies from approximately 25% to 45%. Although heterozygotes are generally asymptomatic, the presence of the abnormal gene can be detected. Some of the erythrocytes in the heterozygote can be made to sickle under laboratory conditions of decreased oxygen tension and provide a means for identifying the presence of the HbS. These individuals are said to possess the *sickle cell trait*.

CHARACTERISTICS. In the homozygote, sickle cell anemia is a severe, chronic anemia beginning toward the end of the first year of life. During the early months of life the child derives some protection from fetal hemoglobin (HbF). This highly specialized hemoglobin with increased oxygen-carrying capacity is produced during fetal life but following birth is gradually replaced by the adult form, or in the sickle cell child by HbS. The initial symptoms of sickle cell anemia usually appear at about 2 to 4 years of age. The child is pale, fatigued, has a poor appetite, and may complain of leg, arm, or back pain. Frequently an acute illness is precipitated by an upper respiratory or gastrointestinal tract infection. The symptoms manifested by an acute attack (sickle cell crisis) are directly related to the areas in which the sickling occurs. Tissue ischemia will result in severe pain in the tissue supplied by the obstructed vessels (Fig. 3-7). Chronic leg ulcers are a frequent finding, and salmonella infections are common. The abdominal pain of sickle cell crisis resembles other acute abdominal emergencies, which confuses the diagnosis in unidentified individuals. Progressive impairment of liver and kidney function is due to repeated attacks, and occlusion of cerebral vessels may be fatal or result in hemiplegia. The condition is sometimes fatal in the first two decades of life, but for those who survive to adulthood the severity of symptoms seems to decrease.

Individuals with the sickle cell trait do not develop the anemia. However, they are less able to withstand some physiologic stresses such as heavy work or excercise. They also are affected by situations involving decreased oxygen content such as high altitudes or a general anesthesia. In these instances there may be intravascular sickling severe enough to cause thrombosis or infarction in the

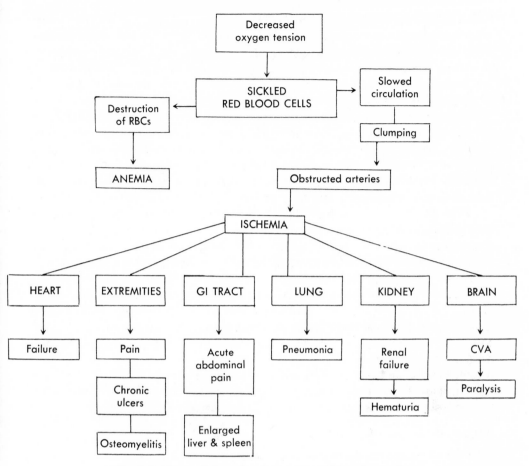

Fig. 3-7. Effects of tissue involvement in sickle cell anemia.

spleen, lungs, or cerebral vessels. Women with sickle cell trait may have more difficulty with pregnancy.

DIAGNOSIS AND TREATMENT. Clinical manifestations are varied and frequently resemble those of other disorders. The positive diagnosis is made by microscopic examination of red blood cells or, more often, electrophoretic examination of the hemoglobin. This method detects both the homozygote and the heterozygote. A relatively quick and simple screening procedure, the Sickledex test, is available for carrier detection. There is no effective treatment for sickle cell anemia; therefore, major efforts are palliative. Analgesics provide relief of discomfort during crises, and dehydration and acidosis must be corrected. Blood transfusions replace blood losses but are of no value during crises.

The thalassemias

These conditions are due to impairment in protein synthesis of hemoglobin rather than an alteration in hemoglobin structure. The thalassemias result from failure to produce a polypeptide chain. There are two major types of thalassemias: α-thalassemia, due to reduced alpha chain synthesis, and β-thalassemia, resulting from defective beta chain production. The ultimate result of these deficits is shortened red cell survival. Both thalassemias occur in a major (homozygous) and a minor (heterozygous) form. Relatively harmless in the heterozygote, they produce severe, almost universally fatal anemias in the homozygous state.

HEREDITY AND INCIDENCE. The thalassemias acquired the name from the Greek word *thalassa,* meaning sea, since most affected persons or their ancestors are from countries bordering the Mediterranean. There is high incidence in Italy and in persons of Italian background but it is also relatively common in Orientals and Africans. The inheritance is described as intermediate since the gene is partially expressed in the heterozygote and fully expressed in the homozygote.

Thalassemia minor. An individual with only one defective gene has sufficient hemoglobin production that the manifestations of the disorder are relatively unremarkable. There is a variable degree of red blood cell fragility and its accompanying mild to serious anemia, especially the β-thalassemia. Some affected individuals have no symptoms.

Thalassemia major. In contrast to the minor forms, thalassemia major produces a severe, fatal anemia. In α-thalassemia major the hemoglobin production is so defective that the homozygote does not survive intrauterine life but is stillborn with hydrops fetalis. The β-thalassemia major (Cooley's anemia) usually develops in early childhood, often during the first year. The child fails to thrive, becomes febrile and progressively anemic. There are also alterations in bone development of the skull. Treatment is palliative, with blood transfusion and sometimes splenectomy. Most afflicted children do not survive to adulthood.

DISORDERS AFFECTING SKELETAL MUSCLES

The numerous hereditary disorders involving skeletal muscles are manifest by muscle weakness and wasting, eventually progressing to atrophy with varying degrees of contracture deformity. The diseases involve either the muscle fibers themselves or some portion of the central nervous system that innervates the affected muscles. Although treatment is palliative and symptomatic rather than curative, an accurate diagnosis is essential for purposes of rehabilitation and counseling. A diagnosis can usually be made on the basis of clinical manifestations and family history, although the mutation rate is relatively high in the X-linked form. In addition, special diagnostic tools such as muscle biopsy, serum enzyme levels, and electromyography are employed to help establish a diagnosis.

The muscular dystrophies

The muscular dystrophies constitute the largest and most important single group of muscle diseases of childhood. They are characterized by progressive weakness and wasting of symmetrical groups of skeletal muscles, with increasing disability and deformity. The several clinical forms of muscular dystrophy have been considered by some to be variations of a single disease; however, the different hereditary mechanisms tend to indicate that different genes are implicated

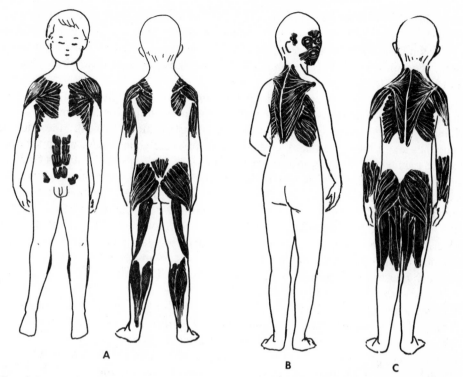

Fig. 3-8. Initial muscle groups involved in the muscular dystrophies. **A,** Pseudohypertrophic; **B,** facioscapulohumeral; **C,** limb-girdle.

in each type. In all forms there is insidious loss of strength, but each differs in regard to muscle groups affected, age of onset, rate of progression, and inheritance patterns. Fig. 3-8 illustrates the muscle groups involved in the three major forms of muscular dystrophy.

The basic defect in muscular dystrophy is unknown although it appears to be due to a metabolic disturbance unrelated to the nervous system. There are elevated serum concentrations of some intracellular enzymes. The most valuable for diagnostic purposes is *creatine phosphokinase* (CPK), which is consistently increased in affected individuals and female carriers of the Duchenne type. This increase in CPK levels affords a means for early detection of asymptomatic children in families at risk, heterozygous carriers, and intrauterine diagnosis of an affected fetus in susceptible pregnancies.

There is no specific therapy for muscular dystrophy. Treatment is mainly confined to supportive measures including physical therapy and other orthopedic procedures to minimize deformity and to assist the affected individual in meeting the demands of daily living.

Pseudohypertrophic (Duchenne). The most severe and the most common muscular dystrophy seen in childhood is the Duchenne type. The inheritance pattern is X-linked recessive and therefore affects males almost exclusively. The onset is early, usually before the fifth year, and relentlessly progresses until death in the second or third decade from cardiac insufficiency or respiratory tract infections. The muscles, involved symmetrically, are those of the pelvic girdle (early) and the shoulder girdle (later) (Fig. 3-8, *A*).

The symptoms may appear in infancy but more often are noted during the preschool years and consist of increasing weakness, difficulty in climbing steps, riding a tricycle, or straightening up from a sitting position. Typically the affected male has a waddling gait and lordosis, falls frequently, and develops a characteristic manner of rising from a squatting or sitting position on the floor (Gower's sign)—he turns onto his side or abdomen, flexes his knees to assume a kneeling position, then with knees extended gradually pushes his torso to an upright position by "walking" his hands up his legs. The muscles, especially the calves, thighs, and upper arms, become enlarged from fatty infiltration, hence the name pseudohypertrophy. Contractures and deformities involving large and small joints are common complications. Ultimately the disease process involves the diaphragm and auxiliary muscles of respiration and cardiac muscle. The cause of death is usually respiratory tract infections or cardiac failure.

Facioscapulohumeral (Landouzy-Dijerine). This milder form of muscular dystrophy usually has its onset in adolescence or early adulthood and initially involves the muscles of the face and shoulder girdle (Fig. 3-8, *B*). The first symptoms are inability to raise the arms above the head and an expressionless masklike face. There may be periods of apparent arrest, but eventually muscles of the pelvic girdle and legs become affected. The inheritance follows an autosomal-dominant pattern, and men and women are equally affected. Progression is very slow and life span is rarely shortened.

Limb-girdle (juvenile muscular atrophy of Erb). Limb-girdle muscular dystrophy has its onset in adolescence or early childhood and is inherited as an autosomal-recessive trait. The first muscles to be affected are usually those of the shoulder girdle but may be the pelvic girdle or both (Fig. 3-8, *C*). The facial muscles are not involved. In boys the disease may be difficult to differentiate from to Duchenne form. The progress and degree of disability varies from slight to severe.

The myotonias

The muscle disorders classified as myotonias are characterized by increased muscular irritability and contractility, with either wasting or hypertrophy of involved muscles. All are aggravated by cold.

Myotonic dystrophy (Steinert's disease). Myotonic dystrophy is one of the more common of the hereditary muscle disorders. The disease is inherited as autosomal-dominant but shows some variability in expression. Symptoms often begin in childhood but may not be apparent until adolescence or well into adulthood, and there is great variation in age of onset within the same family. The muscles primarily involved are the tongue, forearm, and thenar muscles. A prominent effect is that the affected individual has difficulty relaxing his grip. There is often significant weakness of the dorsiflexors of the foot, which interferes with ambulation.

Progressive muscle wasting is the usual course, and lenticular cataracts are almost universal. Other manifestations are frontal baldness (more marked in males) and characteristic progressive emotional and intellectual changes.

Other myotonias. Generalized muscle involvement and hypertrophy of all voluntary muscles are the usual manifestations of *myotonia congenita (Thompsen's disease).* Paramyotonia *(Eulenburg's disease)* is distinguished by tongue myotonus and periodic flaccid weakness of muscle. Both are exacerbated by cold and are inherited as autosomal-dominant. Treatment is symptomatic.

Neurogenic muscular atrophies

This group of disorders affecting skeletal muscles is characterized by progressive muscular weakness and wasting due to degeneration of motor neurons. Most of these diseases are transmitted as autosomal-recessive traits and are usually apparent at birth or shortly after; the earlier the onset, the more rapid the course. The major characteristic is flaccid muscles. Distinction between varieties is not always clear cut.

Infantile spinal muscular atrophy (Werdnig-Hoffman disease). The most severe form of neurogenic muscular atrophy is infantile spinal muscular atrophy, or Werdnig-Hoffman disease. The site of pathology is the anterior horn cells of the spinal cord, but the primary effect is atrophy of skeletal muscles. It is transmitted as an autosomal-recessive trait.

The disorder is manifest early, usually at birth, and inactivity of the infant is the most prominent observation. There is weakness and limited movements of shoulder and arm muscles, and the child lies with his legs in the frog position. Intercostal paralysis results in a collapsed chest and totally diaphragmatic breathing. The cry and cough are weak. Early death from respiratory failure or infection is usual.

Juvenile spinal muscular atrophy (Kugelberg-Welander disease). This disorder is also characterized by anterior horn cell and motor nerve degeneration. Signs and symptoms of proximal muscle weakness appear later, in early childhood or adolescence, and the progression is slower. Many affected persons have a normal life expectancy. There is speculation with some support that this disorder is a variant of the infantile form, perhaps due to incomplete penetrance of the defective gene.

Peroneal muscular atrophy (Charcot-Marie-Tooth disease). There are three different inheritance patterns in peroneal muscular atrophy: autosomal-dominant, recessive, and X-linked. The age of onset is anywhere between 5 and 15 years, with muscle weakness in feet and legs gradually spreading to hands and forearms. Proximal muscles are seldom involved. Progression is slow but inexorable, leading to deformities of hands and feet. The disease tends to be more severe in the autosomal-recessive form, least severe in the dominant form, and intermediate in the X-linked form.

SKELETAL DEFECTS

Heritable skeletal defects are probably numerous but since they are nearly always due to a mutant dominant gene and, except for the more mild deformities, the affected individuals rarely live to transmit the gene, there are few in which a definite mode of inheritance can be determined. It is probable that most of the generalized and many localized skeletal deformities are due to a single gene, although the specific biochemistry has not been determined. A large number of skeletal defects are related to defective development of the embryo from both environmental and genetic causes. Many skeletal abnormalities are part of a recognized metabolic upset or represent one of several manifestations comprising a syndrome.

Variations in digit number and configuration such as *brachydactyly, syndactyly,* and *polydactyly* are relatively harmless in that they do not alter the fitness of the affected person. Some of the more generalized skeletal defects such as *osteogenesis imperfecta* (p. 44) and *arachnodactyly* (p. 44) have been mentioned previously. Multiple exostosis is a disorder characterized by osseous elevations of varying sizes on the bones that cause little difficulty unless pressure symptoms necessitate their removal. Rarely, they may become malignant. All of these disorders are inherited as autosomal-dominant traits and may show marked variation in expression.

Achondroplasia

This relatively common form of dwarfism has been traced through numerous generations in affected families. In achondroplasia there is a defect in ossification at the epiphyseal plate (growth portion of long bones), resulting in very short limbs, large head, and lordosis. A large number of these individuals die before or shortly after birth, but those who survive have good general health and average life expectancy. The disorder is usually transmitted as an autosomal-dominant trait although there is a less common type inherited as autosomal-recessive. Achondroplasia has a relatively high mutation rate (sporadic cases are frequent), and there seems to be a paternal factor since many fathers are appreciably older than average.

DISORDERS OF THE NERVOUS SYSTEM AND SPECIAL SENSE ORGANS

Disorders involving the nervous system are more difficult to attribute to a single gene. So many interrelated factors affect the prenatal development of the nervous system that a mode of inheritance can be identified in only a very few cases. Mental retardation is a frequent manifestation in most chromosome abnormalities (Chapter 4) and in a variety of metabolic disorders due to single genes (PKU, galactosemia, Tay-Sachs disease, and so forth). Single genes are felt to be responsible for at least one form of hydrocephalus (Dandy-Walker syndrome), a form of epilepsy, the neuromuscular disorders described earlier, and a congenital insensitivity to pain. Structural defects of the brain or spinal cord have been attributed to a number of etiologies, and most will be discussed in relation to multifactorial inheritance. The heterogeneity of *hereditary ataxias* (inability to coordinate voluntary muscular movement) creates a similar difficulty in assigning a specific inheritance pattern in a given situation although if the onset is late and direct transmission is demonstrated the pattern is felt to be autosomal-dominant. Two important single-gene disorders of the nervous system are Huntington's chorea (pp. 35 and 186) and neurofibromatosis (p. 161).

Deafness

Heritable hearing disorders are relatively common and nearly always the result of a recessive gene. Because they suffer the same affliction, affected persons tend to seek the company of others similarly affected; therefore, marriages between deaf individuals are frequent. Observation of the offspring of these matings indicates that the responsible genes are not always the same but may be alternative genes at different loci. In a large proportion of marriages between deaf mutes the offspring are affected, as would be expected in autosomal-recessive inheritance. However, a fairly large proportion of such marriages yield normal children. It is apparent in these situations that the parents are not homozygous for the same gene. Sometimes there is a mixture of deaf and normal children, which leads some investigators to suspect the possibility of a dominant trait with variability of expression. A sex-linked form has also been described. In addition, there are known environmental agents that cause deafness (for example, the rubella virus). Such instances serve to illustrate some of the complexities encountered in relation to genetic counseling.

Disorders of vision

Disorders of the eye that are due to a single gene are also numerous and varied. They range from those so mild as to cause very little inconvenience (color-blindness) or moderate difficulty (night blindness, myopia) to those that create severe disability (blindness) or are life-threatening (retinoblastoma). Some ocular disorders occur as isolated defects or may be present as part of a number of syndromes (congenital cataracts and many types of developmental malformations). Only a few disorders will be mentioned.

Errors of refraction. Defects in any or all of the structures comprising the normal components of the eye that are responsible for focusing can result in errors of refraction. Some of these include deviations in the configuration or consistency of the cornea and variations in the depth of the ocular chambers. All the components of refraction vary between individuals so that the largest percentage of refractive errors is probably due to multiple factors. Single-gene factors have been identified in some of the corneal abnormalities and may be transmitted by the mendelian patterns of inheritance.

Myopia, or nearsightedness. This disorder is sometimes due to a recessive gene (although it may occasionally be autosomal-dominant) if environmental factors are ruled out. Diagnosis is usually made when the child is about 3 years of age after parents have observed mannerisms in the child that are characteristic of the nearsighted person. It has a tendency to remain stationary.

Cataracts. Most hereditary cataracts are transmitted as autosomal-dominant but some may be autosomal-recessive. This includes the congenital, juvenile, and senile varieties, although with cataracts in the older age groups it is sometimes difficult to determine the precise etiology. In such cases, relatives are usually affected with the same type of cataract. As mentioned previously, cataracts are frequently associated with other defects or syndromes.

Functional defects. The most common color vision defect is the X-linked variety of colorblindness in which the affected individual (predominantly males) is unable to distinguish one or more colors, for example, red/green color vision loss. Complete color vision loss, which includes day blindness (the ability to see better by night than by day), is autosomal-recessive. Night blindness (marked inability to see at night or in subdued light) can have an autosomal-dominant, recessive, or X-linked inheritance pattern and is usually associated with myopia.

Degenerative disorders. A disorder characterized by progressive degeneration of the retina, *retinitis pigmentosa* occurs with relative frequency. The inheritance pattern can be dominant, recessive, or X-linked, and the disability varies from slight to severe. The most common and most severe form is the autosomal-recessive disorder, with a frequency of approximately 1 in 4,000.

REFERENCE

1. Macalpine, I., and Hunter, R.: Porphyria and King George III, Sci. Am., July 1969.

GENERAL REFERENCES

Beutler, E.: Genetic disorders of red cell metabolism, Pediatr. Clin. North Am. **53**:813, 1969.

Brady, R. D.: Genetics and the sphingolipidoses, Med. Clin. North Am. **53**:827, 1969.

Brinkhous, K. M.: Changing prospects for children with hemophilia, Children, **17**: 222, 1970.

Clark, C. A., editor: Selected topics in medical genetics, London, 1969, Oxford University Press, Inc.

Cohn, H. D.: Hemostasis and blood coagulation, Am. J. Nurs. **65**:116, 1965.

Dean, G.: The porphyrias, Br. Med. Bull. **25**: 48, 1969.

Frezal, J., and Rey, J.: Genetics of disorders of intestinal digestion and absorption. In Harris, H., and Hirschhorn, K., editors: Advances in human genetics, New York, 1970, Plenum Publishing Corp.

Guyton, A. C.: Textbook of medical physiology, ed. 4, Philadelphia, 1971, W. B. Saunders Co.

Hirschhorn, K.: Errors of metabolism in children, Hosp. Med. **5**:77, 1969.

Knox, W. E.: Inborn errors of metabolism. In Fishbein, M., editor: Birth defects, Philadelphia, 1963, J. B. Lippincott Co.

Nelson, W. E., Vaughn, V. C., and McKay, R. J.: Textbook of pediatrics, ed. 9, Philadelphia, 1969, W. B. Saunders Co.

Pochedly, C.: Sickle cell anemia; recognition and management, Am. J. Nurs. **71**:1948, 1971.

Porter, I. H.: Heredity and disease, New York, 1968, The Blakiston Division, McGraw-Hill Book Co.

Roberts, J. A. F.: An introduction to medical genetics, ed. 5, London, 1970, Oxford University Press, Inc.

Scott, C. I., and Thomas, G. H.: Genetic disorders associated with mental retardation, Pediatr. Clin. North Am. **20**:121, 1973.

Stanbury, J. B., Wyngaarden, J. B., and Fredrickson, D. S., editors: The metabolic basis of inherited disease, ed. 3, New York, 1972, McGraw-Hill Book Co.

Strauss, H. S.: Diagnosis of inherited bleeding disorders, Pediatr. Clin. North Am. **19**:1009, 1972.

Whissell, D. Y.: Hemophilia in a woman, Am. J. Med. **38**:119, 1965.

4

Chromosomal aberrations

An aberration is defined as a deviation from that which is normal or typical. Aberrations of chromosomes are usually deviations in either structure or number, and the consequences in either situation can be readily observed in the phenotype. Since the development of improved techniques for the study of chromosomes, more physical disabilities have been recognized as having their origin in chromosome defects. It is estimated that approximately 1% of anomalies in newborn infants can be attributed to chromosome abnormalities. Although the types of chromosome disorders are not as varied as those due to a single mutant gene, the incidence for many of the specific abnormalities is significantly higher than single-gene disorders. For instance, Down's syndrome, the chromosome anomaly encountered most frequently, occurs once in every 600 to 700 births, and identification of chromosome abnormalities in early abortuses indicates that they are probably responsible for a large percentage of fetal wastage.

TYPES OF CHROMOSOME ABERRATIONS

The complex nature of cell division makes it highly susceptible to mechanical error, particularly during the critical processes of gamete formation and the early divisions in the zygote (see Chapter 1). The bulk of abnormalities accounted for in the literature is related to variations in chromosome number; however, there is evidence that with technical advances the structural abnormalities will assume greater significance in relation to human disease.

Aberrations in chromosome structure

Occasionally an alteration occurs in normal chromosome structure. It may involve loss or rearrangement of the genes of a single chromosome or the exchange of genetic material between nonhomologous chromosomes and will produce adverse phenotypic effects with varying degrees of severity. Some of the major structural

73

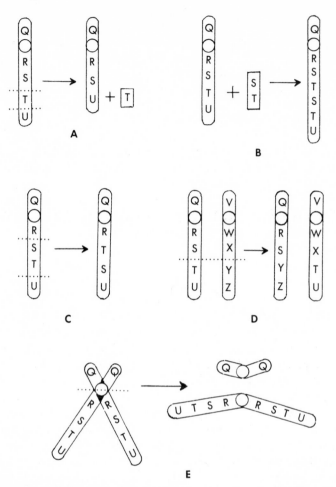

Fig. 4-1. Major types of aberrations in chromosome structure. **A,** Deletion; **B,** duplication; **C,** inversion; **D,** translocation; **E,** isochromosome.

abnormalities (Fig. 4-1) include: (1) *deletion,* in which a portion of a chromosome and the genes within the missing segment are lost; (2) *duplication,* in which a deleted portion of a chromosome becomes attached to another chromosome with a normal gene complement; (3) *inversion,* in which the linear arrangement of genes in a chromosome segment is broken and the order is reversed when the segment is reattached; (4) *translocation,* in which there is an exchange of parts between two nonhomologous chromosomes; and (5) formation of *isochromosomes,* in which there is transverse rather than longitudinal division at the centromere to form a chromosome whose arms are identical in length and gene composition.

Aberrations in chromosome number

Deviations in chromosome number are broadly classified in terms of loss or gain of chromosome *sets.* They may be of two types: euploidy or aneuploidy.

Euploidy. When the alteration in chromosome number is balanced, that is, an exact multiple of the basic chromosome number, it is *euploid.* For example, in man the normal chromosome number is 46 (23 pairs). Cells with a single set of 23 chromosomes (the normal complement of gametes) are *haploid* (n), or *monoploid.* Somatic cells with two sets are *diploid* (2n or 46). Aberrant cells that contain three, four, or more complete sets of chromosomes are *triploid* (3n or 69), *tetraploid* (4n or 92), and so on. The term *polyploid* is often used to describe any multiple greater than the two haploid sets found in normal cells.

Aneuploidy. A chromosome complement that is abnormal in number but is not a multiple of the haploid set is *aneuploid.* Numerical alterations in individual chromosomes are designated with the suffix -*somy.* For example, a cell that contains one less than the basic number is a *monosomy,* or *monosomic* (2n – 1 or 45), because one chromosome pair has but a single member. Cells that bear three identical chromosomes instead of a normal pair are *trisomic* (2n + 1 or 47); cells with more than three are *polysomic.*

Mosaicism. All the body cells may contain the variation in chromosome number, or only a portion of the cells may be aneuploid. That is, some cells will have a normal number of chromosomes and some will have an abnormal number of chromosomes—a mixture of normal and trisomic cells. This situation is termed *mosaicism* and occurs more frequently in relation to abnormalities involving the X chromosome.

MECHANISMS PRODUCING CHROMOSOME ABERRATIONS

The mechanism responsible for producing polyploidy is thought to be a failure of cytokinesis following nuclear division. The chromosomes duplicate, but a new membrane fails to synthesize and separate the cytoplasm to form two daughter cells. Total polyploidy appears to be lethal in man and is evident in many early abortuses; however, polyploid cells are found in some tissues such as liver, bone marrow, and neoplasms.

A significant number of aneuploidies occur in man that are compatible with life, especially those involving the sex chromosomes. More serious outcomes are related to abnormalities of the autosomes. Apparently all chromosome material is so essential to life that loss of even a small chromosome has lethal consequences to the organism. There has been no reported instance of an individual with monosomy of an autosome; trisomies are the chromosome aberrations encountered most frequently by health personnel. The mechanisms usually considered to be responsible for unequal distribution of genetic material are *nondisjunction* and *anaphase lag;* less frequent causes are *translocation* and *chimerism.*

Nondisjunction

Disjunction refers to the separation and migration of chromosomes during cell division; failure of this process is termed *nondisjunction.* The consequence of this prolonged attachment is an unequal distribution of chromosomes between the two daughter cells (Fig. 4-2, *A*). *Anaphase lag* refers to the failure of a chromosome to migrate after centromere-splitting (Fig. 4-2, *B*). In this case half of the chro-

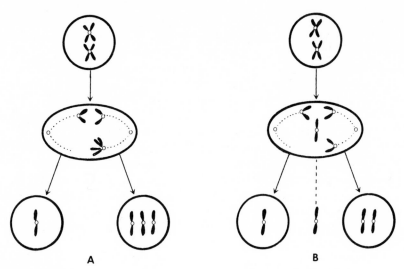

Fig. 4-2. Mechanisms of maldistribution of chromosomes. **A,** Nondisjunction; **B,** anaphase lag.

mosome pair remains in the center of the dividing cell and either wanders to the opposite pole to form a trisomy or is lost during division of the cytoplasm. These accidents have similar outcomes and can occur before fertilization, during either of the meiotic divisions of germ cell formation, or after fertilization during early mitosis in the zygote. For convenience, the term nondisjunction will be used to describe these errors of chromosome movement and their consequences.

The factors responsible for nondisjunction seem to be related to spindle activity, but the precise nature of the defect is unclear. There is some relationship between chromosomes that are late-replicating* and the frequency of nondisjunction. This is especially true of the sex chromosomes and is reflected by the significant majority of aneuploidies involving the X or Y chromosomes. There is a definite correlation between some chromosomal aberrations and advancing maternal age, which is well established in regard to the increased incidence of trisomy 21, or Down's syndrome, in children of mothers over 35 years of age. The possibility also exists that there may be a genetic susceptibility to nondisjunction. There are some studies that implicate other factors such as chemicals, radiation, and viruses as possible etiologic agents but the evidence is not conclusive.

Nondisjunction during meiosis. Fig. 4-3, *A* illustrates the maldistribution of chromosomes that results from nondisjunction in each of the meiotic divisions and the zygotes formed at fertilization. Only the activity of chromosomes in the female germ cells are described, and the polar bodies are omitted to better indicate the four possible types of gametes. Nondisjunction can take place during oogenesis or spermatogenesis (rare) and can involve autosomes or sex chromosomes.

---

*Chromosomes separate at different rates during anaphase. Those which tend to be slower in pulling apart are late-replicating chromosomes.

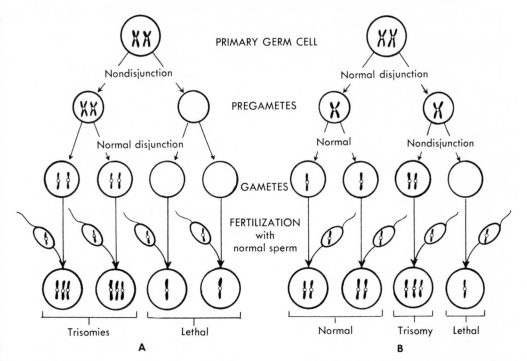

Fig. 4-3. Mechanisms of maldistribution of chromosomes during meiosis in the ovum and fertilization with normal sperm. **A,** During first meiotic division; **B,** during second meiotic division.

Nondisjunction during the first meiotic division occurs when a synapsed chromosome pair remains together and migrates to the same pole. As a result, half the gametes will contain 22 chromosomes and half will contain 24. Any mature gamete produced from this one germ cell will contain an abnormal chromosome complement and upon union with a normal gamete will produce either a monosomic or a trisomic cell.

Failure of the chromosome to split during the second meiotic division results in a similar maldistribution. Since the mechanism probably involves only one of the pregametes (secondary oocyte or spermatocyte), two of the resulting gametes will contain a normal chromosome complement and two will be aneuploid. Union with normal gametes will form, respectively, one monosomic and one trisomic cell (Fig. 4-3, *B*). The associated clinical anomalies seen in the phenotypes of the above kinds of aneuploids will be discussed later in this section. It is possible for nondisjunction to take place during both of the meiotic divisions and is one possible explanation for the rare individual with supernumerary sex chromosomes.

Nondisjunction during mitosis. When nondisjunction occurs during early mitotic division following fertilization, the results are similar to those of meiotic nondisjunction. The principle difference lies in the overall chromosomal constitution of the individual. Whereas nondisjunction before fertilization produces a zygote in

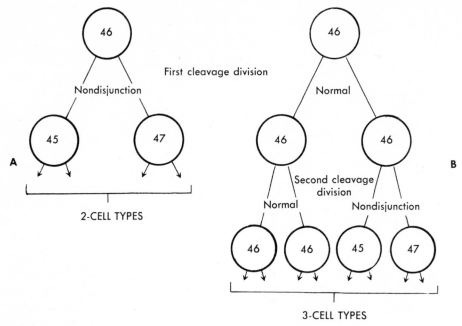

Fig. 4-4. Nondisjunction in early mitotic division of the zygote, which produces a mosaic genotype. **A,** Disjunction at first mitotic division; **B,** disjunction during the second mitotic division.

which all the cells contain the abnormal chromosome complement, nondisjunction after fertilization results in two or more distinct cell types. Mitotic, or postzygotic, nondisjunction can involve either autosomes or sex chromosomes and is the mechanism responsible for mosaicism. Nondisjunction during the first mitotic division (cleavage) of the zygote produces two cell types: monosomic (N = 45) and trisomic (N = 47). Nondisjunction during the second cleavage will produce three cell types: one half normal and the other half an equal distribution of monosomic and trisomic cells (Fig. 4-4). Monosomic cells (with the exception of the X monosomy) are nonviable so that most mosaics are an intermixture of normal and trisomic cells.

Translocation

Translocation is a defect in chromosome structure rather than in cell division. It involves the transfer of genetic material from one chromosome to a nonhomologous chromosome. The most frequently encountered translocations are between acrocentric chromosomes, the best known being the fusion of a D and a G chromosome as seen in familial Down's syndrome. Whereas nondisjunction is usually a sporadic event, translocation is more often hereditary and occurs in both males and females.

The individual with a translocated D/G chromosome is phenotypically normal. His cells contain the normal amount of genetic material although it is contained in

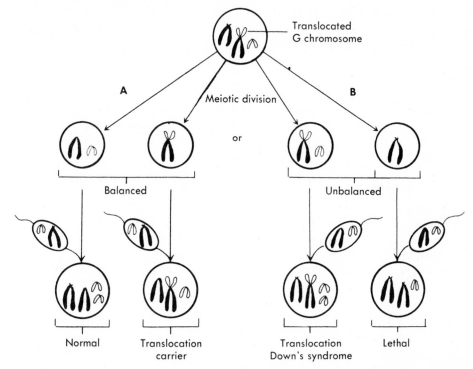

Fig. 4-5. Normal cell division of a translocated chromosome. **A,** The hereditary material is divided equally between the resulting gametes. **B,** The hereditary material is unequally divided in the gametes.

only 45 chromosomes. When the long arms of the D and G chromosomes become attached, they give the appearance of one large submetacentric chromosome and there is one less chromosome in the G group. Gametogenesis involving the translocated chromosome may proceed in two directions to yield two different types of gametes (Fig. 4-5). When the translocated chromosomes are distributed to one gamete and their unaffected homologues are distributed to the other, the gametes are said to have *balanced* chromosomal constitutions. If the translocated chromosomes and a homologue of either the D or the G chromosome are distributed to the same gamete, it is said to be *unbalanced:* there is unequal distribution of the D and G chromosome material.

The balanced division produces zygotes that are phenotypically normal and contain either a normal chromosome complement (N = 46) or the translocated chromosome plus the remaining 44 (N = 45). The gametes with the unbalanced chromosome complement will produce two zygotes: one that lacks a chromosome (N = 45) and one with a normal number (N = 46) but which contains the genetic material of one chromosome in triplicate. Thus, a translocation carrier mated to a normal individual can be expected to produce four types of offspring: normal individuals, viable defective children, normal translocation carriers, and nonviable monosomies (Fig. 4-5). Four types of gametes can be produced and fertilized; however, since the nonviable zygote is aborted early, the chance of producing an affected child is actually one in three.*

Chimerism

Chimerism is a rare phenomenon that is occasionally encountered in a pair of fraternal (nonidentical) twins. During intrauterine life, blood cells may be exchanged by transplacental transfusion from one twin to the other. The transplanted blood cells continue to survive and proliferate so that the host will contain blood cells of two different genetic types.

CONDITIONS DUE TO AUTOSOMAL ABNORMALITIES

Autosomal abnormalities are the basis for many disorders and anomalies and most frequently involve the late-replicating chromosomes 13, 18, and 21. It is also of interest that the effects of trisomies seen clinically seem to involve only the small chromosomes; there has been no reported trisomy for any of the large chromosomes. The autosomal trisomy often encountered by workers in health-related and education fields is Down's syndrome: a trisomy of one of the smallest chromosomes, number 21. Trisomies of two other chromosomes, number 13 and number 18, are seen less frequently. In all three the total chromosome count is 47 (except for the translocation in which the number is 46).

The modified standardization of chromosome nomenclature agreed upon at the Chicago Conference (1966)[1] identifies individual chromosome abnormalities according to chromosome number, sex, and the addition or deletion of the specific chromosome involved. For example, the complement is recorded as 46,XY or 46,XX for a normal male and female, respectively. Numerical aberration such as a male with an extra G chromosome is described as 47,XY,G+. The precise chromosome is indicated by number when it can be identified, for example, 47,XY,21+. Structural alterations are designated by the use of several lower-case symbols. The short arm of a chromosome is designated p, the long arm q, and a translocation by the letter t. A translocation between a D and a G group chromosome would be indicated as 45,XX,D–,G–,t(DqGq)+, which signifies that there are a total of 45 chromosomes, XX sex chromosomes, a missing chromosome from the D group and one from the G group, with their long arms uniting to form a D/G chromosome. The broad classification is used to describe the disorders presented in the following sections. The extra chromosome is designated as a trisomy, with the involved group and specific chromosome indicated.

Trisomy 21 syndrome

Trisomy 21 is variously known as trisomy G, "mongolism," or Down's syndrome, and is a consequence of nondisjunction of one of the G-group chromosomes, probably number 21 (Fig. 4-6). The term mongolism was assigned to this syn-

---

*Theoretically, six types of gametes can be produced depending on whether it is the D or G chromosome that travels with the translocated chromosome.

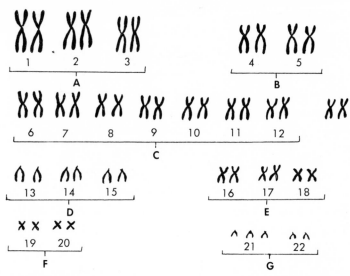

Fig. 4-6. Model of a female karyotype with trisomy of the 21 (G) chromosome group (47,XX,21+).

drome by Dr. J. H. Langdon-Down, who interpreted the peculiar facial characteristics as a "reversion to mongoloid stock." It is both inaccurate and distasteful but
continues to be used by both lay and professional people. Although trisomy 21 is
more descriptive, the term Down's syndrome is gaining wider usage and will be
used in this discussion.

HEREDITY AND INCIDENCE. In approximately 95% of cases, Down's syndrome
can be attributed to meiotic nondisjunction, usually in the maternal gamete. The
relationship between Down's syndrome and advancing maternal age has been well
demonstrated: the frequency in young mothers is approximately 1 in 1,000 to
2,000 births; at the age of 35 years the incidence increases to 1 in 300; from the
age of 40 the incidence rises sharply to become 1 in 30 to 50 in the over-45 age
group (Fig. 4-7). This relationship was made by very early investigators when
they observed Down's syndrome frequently affected the last child born in a family.
The number of previous pregnancies is not a factor, nor is there an independent
paternal age effect. The basis for these observations is that there is a continuous
production of fresh sperm but the eggs in the adult ovary have remained at
prophase I since intrauterine life. In addition to the effects of their increased age,
at each ovulation the germ cells must proceed through both meiotic divisions,
which doubles the chance for mechanical error. Women over 30 are old from a
cytogenetic point of view.

A small percentage (5) of cases are due to translocation in which the 21
chromosome becomes attached to another chromosome, usually one of the D
group (or occasionally to the number 22). In this situation maternal age is no
factor. Parents are generally in the younger age group, and often one parent is
found to be a carrier with a normal phenotype and a chromosome complement

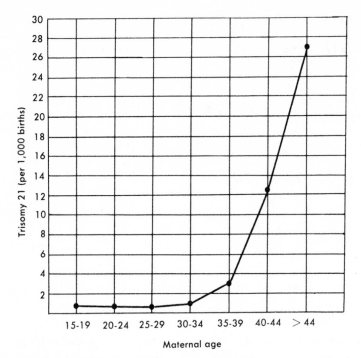

Fig. 4-7. The accelerating rate of increase of Down's syndrome in the offspring of mothers over the age of 30. (From Reisman, L. E., and Matheny, A. P.: Genetics and counseling in medical practice, St. Louis, 1969, The C. V. Mosby Co.)

of 45. As is evident from the discussion of translocation, the recurrence risk in this form of Down's syndrome is significantly higher than it is in the nondisjunction variety; theoretically, one third of live offspring will be affected. Identification of a carrier, especially in young parents, is of vital concern for purposes of genetic counseling.

Occasionally a person with Down's syndrome will display both trisomic and normal cell lines, indicating a nondisjunction during early mitotic division in the zygote. This is mosaicism, and although usually of the 46/47 combination there may be other variants. Clinical manifestations in this form are fewer, depending upon the tissues involved, and may show considerable variation. This is especially true in relation to the degree of mental deficiency, which may be extremely severe or rather mild.

DIAGNOSIS. Down's syndrome is easily recognized, and the diagnosis can nearly always be made early on physical characteristics alone. The most consistent observable findings associated with Down's syndrome are: short stature; small, round head with small, low-set ears and flattened occiput; epicanthal folds and oblique palpebral fissures (mongoloid slant of the eye); small mouth with protruding tongue; hypotonic (flabby) muscles and hypermobility of the joints; broad, short hands with stubby fingers, inward curved little finger, and a transverse palmar crease; and, the most significant feature, mental retardation. The diagnosis is usually made at birth on the basis of the total picture, and almost always before the age of 6 months. In doubtful cases, analysis of dermatoglyphic (handprint) patterns can often confirm the diagnosis. In addition to the transverse palmar crease or so-called simian line, persons with Down's syndrome have characteristic dermal ridge patterns (Appendix C). Positive confirmation is made on the basis of chromosome analysis.

There are a number of major anomalies and illnesses that are frequently associated with this disorder. Approximately one half of the children with Down's syndrome have congenital heart disease, which is probably responsible for a large percentage of mortality in the first year of life. Other structural defects with increased incidence in Down's syndrome are renal agenesis and gastrointestinal defects such as duodenal atresia and aganglionic megacolon (Hirschsprung's disease). There is also a significant increase in the incidence of leukemia in children with Down's syndrome.

TREATMENT. There is no cure for Down's syndrome. Surgical correction of heart and other structural defects, as well as availability of antibiotics to combat upper respiratory diseases to which they seem particularly susceptible, have significantly increased the life expectancy of persons with this disorder. The chief disability is the marked degree of mental retardation, which causes the affected individual to be dependent upon others for his care and existence. This disorder accounts for approximately 10% of institutionalized mentally defective people. The major effort of those concerned with counseling and guidance is directed toward assisting anguished parents to cope with the burdens that such a child (or adult) places upon them. As a rule, the person with Down's syndrome is relatively easy to care for compared with other mentally deficient individuals; he is normally pleasant and cooperative, and his social development is quite satisfactory. In a few persons with the mosaic variety of Down's syndrome, in which some systems contain normal cells, IQ is near the low-normal range.

Persons with Down's syndrome rarely reproduce. The males are apparently sterile; but some females, although they have a late menarche, scanty irregular menstrual periods, and an early menopause, have borne children. In such instances there is a 50% chance that the offspring will be similarly affected.

Trisomy 18 syndrome

Trisomy 18, or *Edward's syndrome,* is seen less frequently than trisomy 21 but more often that trisomy 13. Trisomy 18 occurs in approximately 1 per 5,000 births and in three times as many females as males. Characteristic features are micrognathia, deformed and low-set ears, rocker-bottom feet, overlapping of fingers (index over third finger), prominent occiput, wide-spaced eyes, Meckel's diverticulum, and mental retardation. There are usually associated cardiac and renal anomalies. The facial features and the peculiar finger abnormality make this syndrome easily recognized at birth. Failure to thrive and early death are usual.

Trisomy 13 syndrome

The physical anomalies associated with trisomy 13, or *Patau's syndrome,* are so numerous and severe that afflicted babies rarely survive beyond early infancy.

The multiple abnormalities seen most frequently include cleft palate, harelip (often bilateral), ear malformations, multiple hemangiomas, microphthalmos, polydactyly, eye defects, and mental retardation. The incidence is approximately 1 in 15,000 births, the great majority being female.

Cri du chat (cat cry) syndrome

The chromosome defect responsible for this disorder is deletion of the short arm of one chromosome in the B group, probably number 5. The syndrome derived its name *(la maladie cri du chat)* from the distinctive weak, high-pitched, mewlike cry resembling that of a cat. Other clinical features include a small head, wide-spaced eyes, often with epicanthal folds, failure to thrive, and profound mental retardation.

SEX CHROMOSOME CHARACTERISTICS

Like those of the autosomes, sex chromosome aberrations are primarily due to maldistribution of the genetic material during gametogenesis or early cleavage division in the zygote, and there are a number of conditions due to duplicated or missing sex chromosomes. Observation and research into these disorders have helped in better understanding of the function of sex chromosomes in normal persons.

Sex determination

Sex is genetically determined at the time of fertilization. All female cells contain only one kind of sex chromosome, the X chromosome, which pair and segregate during meiosis to produce only X-bearing gametes (homogametic). Male cells have both X and Y chromosomes (heterogametic) (p. 14). The genetic sex of the offspring depends entirely upon whether the male parent contributes an X or a Y chromosome. Methods have been advanced, whose proponents claim are successful, whereby man can willfully control the sex of his offspring; however, they are not endorsed by those considered to be experts in the field.

Sex chromosome function

The presence of a Y chromosome seems to be essential for the development of testes, their hormones, and eventually the male phenotype. Whenever a Y chromosome is present, the phenotype will be male even in individuals whose cells have been found to contain more than one X chromosome. More than one Y chromosome does not increase "maleness," but the presence of at least one X chromosome does seem to be essential to survival. There are females with only one X chromosome (XO), and there is a high incidence of this anomaly in abortuses. Although theoretically possible, there have been no reported cases of males with a Y chromosome and no X chromosome (YO) or individuals without sex chromosomes (OO), even in abortuses.

Sex chromatin

Unlike the autosomes, an increase in the number of sex chromosomes does not produce the profound effects that are associated with the autosomal trisomies,

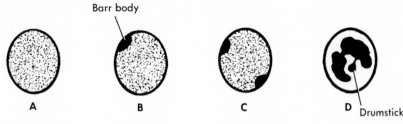

Fig. 4-8. Sex chromatin, or Barr body. **A,** No sex chromatin is found in normal male somatic cells. **B,** One Barr body is normal in female somatic cells. **C,** Two Barr bodies are found in cells with three X chromosomes (XXX or XXXY). **D,** The drumstick is found in many polymorphonuclear leukocytes of the normal female.

although some degree of mental deficiency accompanies many of them and the affected individuals are with few exceptions sterile. In fact, many are not recognized until they seek consultation for infertility. An explanation for this reduced disability in individuals with multiple sex chromosomes is based on the X inactivation concept (Lyon hypothesis) discussed in Chapter 2. Because it is nonfunctioning, this X chromosome exerts little effect on the phenotype. However, this inactivated X chromosome, represented by the sex chromatin (Barr) body, provides a simple means for detecting its presence or absence in the interphase nuclei of somatic cells (Fig. 4-8). It has been established that the number of sex chromatin bodies is one less than the total number of X chromosomes in that nucleus; therefore, female somatic nuclei are normally *chromatin-positive* (contain one active and one inactive X chromosome), and male cells are *chromatin-negative* (contain only one X chromosome). Females who are chromatin-negative have been found to have the genotype XO, and chromatin-positive males have an XXY genotype. In some polymorphonuclear leukocytes, the chromatin body takes the form of nuclear appendages called drumsticks; the number of drumsticks, again, is one less than the number of X chromosomes (Fig. 4-8, *D*).

The sex chromatin test provides a simple and convenient method for determining genetic sex or for detecting the presence or absence of sex chromosomes in man. Cells from any tissue can be used for this test, but a buccal smear is a cheap and easily accessible technique. Mucosal cells, gently scraped from the inside of the cheek, are spread on a clean glass slide, prepared, and stained. These chromatin bodies can be seen in approximately 20% to 60% of the nuclei of normal females. The sex chromatin test is also proving of value in detecting possible genetically determined disorders in the prenatal period. For example, fetal cells from amniotic fluid are sometimes used to determine the sex of an unborn child in cases where the mother is a carrier of an X-linked disorder.

CONDITIONS DUE TO SEX CHROMOSOME ABNORMALITIES

Compared with most hereditary disorders, sex chromosome aberrations are encountered with relatively high frequency. The possible mechanisms by which they may occur are those previously described; that is, prefertilization nondisjunction during one of the meiotic divisions of gametogenesis in either parent or in the

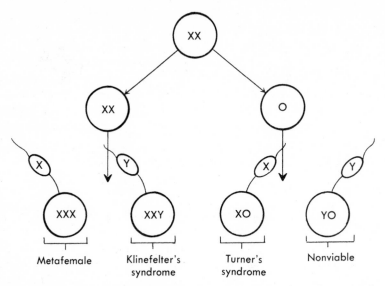

Fig. 4-9. Nondisjunction of X chromosomes in the ovum fertilized by normal sperm to produce the more common sex chromosome aberrations.

early postfertilization (mitotic) divisions of the zygote (Fig. 4-9). Although evidence is not conclusive, there is reason to believe that the X-Y association during spermatogenesis may be responsible for producing abnormal gametes. These nonhomologous chromosomes replicate later than the other chromosomes and tend to form somewhat loose end-to-end associations rather than the usual side-to-side synapsis characteristic of homologous chromosomes. Both of these peculiarities predispose the sex chromosomes to nondisjunction.

Most sex chromosome aberrations are due to an increase in sex chromosome number as a result of nondisjunction during meiosis. Unlike the autosomal abnormalities, a variety of sex chromosome mosaics are known to result from postfertilization nondisjunction. Most of the sex chromosome disorders are identified according to chromosome constitution, as determined by sex chromatin or cytologic analysis. The most common method for describing the disorders is to indicate the total number of chromosomes, followed by the sex chromosome complement. For example, a normal male is 46,XY and a normal female is 46,XX. Some authors, for simplicity, often refer to the sex chromosomes only; for example, an XXX female rather than 47,XXX.

Klinefelter's syndrome

The most common abnormality in the male (and probably of all sex chromosome anomalies) is the 47,XXY genotype also known as Klinefelter's syndrome and chromatin-positive testicular tubular dysgenesis. Although the majority are of the XXY genotype, there are numerous variants, such as XXYY, XXXY, XXXXY, and XXXXYY. The clinical features are essentially the same in all;

however, the degree of mental retardation has a direct relationship to the number of X chromosomes in the genotype. As a rule, the severity of retardation is directly related to an increased number of X chromosomes. Also, the male phenotype seems to be dependent upon the presence of the Y chromosome regardless of the number of X chromosomes; whenever there is a Y chromosome, the phenotype is male in all sex chromosome abnormalities.

HEREDITY AND INCIDENCE. Klinefelter's syndrome occurs in approximately 1 in every 500 male births, probably as a consequence of meiotic nondisjunction during gametogenesis.

DIAGNOSIS. The presence of this disorder is usually not apparent from physical characteristics until the time of puberty. These boys tend to be tall and eunuchoid, with sparse pubic and facial hair. They frequently develop enlarged breasts (gynecomastia), and the testes are small, firm, and insensitive. Spermatogenesis is absent, but sexual urge, although weak, is present, and these males can engage in sexual activities. Mental retardation of varying degrees is a frequent finding, but many affected persons have normal intelligence. All the symptoms are more severe when the number of X chromosomes is greater than two. The diagnosis is usually made on the basis of these clinical features and a chromatin-positive buccal smear.

TREATMENT. The main effort in medical treatment is directed toward enhancing the masculine characteristics through the administration of male hormones, principally testosterone. Gynecomastia may require surgical correction. The psychologic factors associated with the physical features and sterility may have implications for counseling and guidance.

Turner's syndrome

The terms Turner's syndrome and *gonadal dysgenesis* are applied to phenotypic females with the genotype 45,XO.

HEREDITY AND INCIDENCE. Turner's syndrome occurs in approximately 1 in 1,500 to 3,000 live female births and is probably due to meiotic nondisjunction. Unlike most other chromosome anomalies, Turner's syndrome has no apparent relationship to maternal age; in fact, there is evidence to indicate that it might be due to meiotic errors on the paternal side.

DIAGNOSIS. The characteristic features associated with Turner's syndrome are short stature (many are under 5 feet tall), low posterior hairline, webbed neck, and a shieldlike chest with widely spaced nipples. The genitalia are hypoplastic, and there is absence of sexual development at puberty, with primary amenorrhea and sterility. Heart defects, especially coarctation of the aorta, are often associated with this disorder, and there is sometimes a moderate degree of mental retardation although some feel this may be a learning difficulty rather than a mental deficit. The adult with Turner's syndrome is timid and dependent, with generally childlike behavior. The disorder can be recognized in infancy by the characteristic webbed neck, low hairline, and a typical peripheral lymphedema of the lower legs and forearms. Diagnosis is confirmed by a negative sex chromatin; cytologic analysis is rarely necessary.

TREATMENT. Secondary sex characteristics can be produced with estrogen therapy and may even be accompanied by personality changes; there is, however, no effect upon sterility.

XXX female

Trisomy of the X chromosome occurs with greater frequency than the X monosomy, and the incidence is estimated at 1 in 1,000 female births. Females with this 47,XXX genotype are sometimes known as *metafemales* or even *superfemales* and do not differ markedly from normal. Although they are often fertile, they usually have irregular menstruation and early menopause. The incidence of mental deficiency is apparently increased in these women. Sex chromatin tests indicate two Barr bodies.

Based on chromosome activity during meiosis it would be expected that half the gametes and thus half the offspring of these women would be similarly affected. This does not seem to be the case, the theory being that there is some selection against the extra X that causes it to migrate to the polar body during meiosis.

XYY male

There is a great deal of speculation regarding this 47,XYY anomaly since it was discovered to have a higher incidence in individuals institutionalized because of subnormal intelligence and violent, antisocial behavior. These men tend to be tall (usually over 6 feet) and there is no effect on fertility. The extra Y chromosome does not produce additive male effects. So far, there has been insufficient data to confirm the actual incidence in the population or the behavioral characteristics attributed to this genotype.

Mosaics

A number of mosaic genotypes have been reported with considerable variability in phenotypes, including XO/XX, XO/XXX, XO/XX/XXX with female phenotype; XY/XXY, the most common mosaic with a male phenotype (a variant of Klinefelter's syndrome); and XO/XY, XX/XXY, among those in whom the phenotype may show considerable variation. A large percentage of the so-called *intersex* conditions, in which the external genitalia are ambiguous, are the XO/XY genotype, but sex chromatin tests in any of the intersex conditions are inconclusive; karyotyping is the only means by which genetic sex can be established. The problem of intersex will be considered again in Chapter 5 in relation to defects in development.

REFERENCE

1. Chicago conference: Standardization in human cytogenetics. Birth defects, original article series **2:**2, New York, 1966, The National Foundation.

GENERAL REFERENCES

Bartalos, M., and Baramki, T. A.: Medical cytogenetics, Baltimore, 1967, The Williams & Wilkins Co.

Eggen, R. R.: Chromosome diagnostics in clinical medicine, Springfield, Ill., 1965, Charles C Thomas, Publisher.

Ford, C. E.: Mosaics and chimeras, Br. Med. Bull. **25:**104, 1969.

McKusick, V. A.: Human genetics, ed. 3, Englewood Cliffs, N. J., 1969, Prentice-Hall, Inc.

Neu, R. L., and Lytt, I. G.: Clinical aspects

of abnormalities of the X and Y chromosomes, Clin. Obstet. Gynecol. **15:**141, 1972.

Scott, C. I., and Thomas, G. H.: Genetic disorders associated with mental retardation, Pediatr. Clin. North Am. **20:**121, 1973.

5

Genetic and environmental influences on differentiation and development

The formation of a new individual is one of the most dramatic events known to man. Each person's genetic potential is determined at the time of conception when the mature sperm with its genetic material penetrates the mature ovum with its homologous complement of genes to produce a single cell with the combined potentials of both parent organisms. Once fertilization takes place the cell, or zygote, immediately begins the immensely complex processes of growth, development, and differentiation. From this single cell will evolve millions of cells with the diversity to carry on an enormous variety of essential functions. Regulation of these processes is controlled by gene action, each exerting its effect in appropriate tissues at appropriate times. No less significant are the influences of environment, particularly during the time of critical differentiation. This chapter is concerned with the role played by genes in growth and development of the human organism and some of the environmental factors that can alter the normal course of events.

DEVELOPMENT AND DIFFERENTIATION

When he begins his existence a human being bears no resemblance to the complex organism into which he will develop. In fact, during the very early stages he is indistinguishable from any other animal species. The cell, with a nucleus and cytoplasm, contains no structures that remotely correspond to any of the organs and tissues that go to make up the fully developed individual. Development from zygote to adult organism consists of two distinct but interrelated processes: growth and differentiation. Growth results when cells divide and syn-

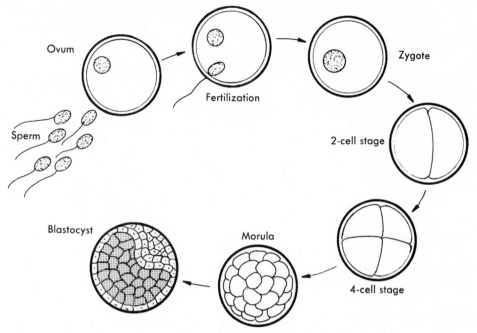

Fig. 5-1. Fertilization and early cleavage division of the zygote.

thesize new proteins and is reflected in increased size and weight. Growth alone would produce a very large mass of similar cells instead of the complex adult with large numbers of cells with diverse and distinct functions. Differentiation is the process by which these early cells are systematically modified and specialized to form all the tissues that are necessary to assure an organized, coordinated individual.

Embryogenesis

Soon after the stimulation of fertilization the ovum begins to divide with orderly precision (Fig. 5-1). The zygote divides by *cleavage* into two smaller, similar cells called *blastomeres*. The blastomeres divide into four, then into eight, and so on, to form a solid ball of cells called a *morula* (so named because it bears a resemblance to a mulberry; Lat. *morus,* mulberry). In man, cleavage divisions are always mitotic; therefore, the daughter cells receive the full complement of chromosomes. These cells do not increase in size between divisions so that, although the cells increase in number, there is little if any increase in the size of the early developing embryo from that of the original zygote. At this stage the cells are simple structures that are fairly uniform in size, shape, and physiologic capabilities. If the cells are separated during these very early divisions (for example, the two-cell or four-cell stage) each will develop into a complete embryo.

Embryonic cells undergo a number of divisions before they begin to differentiate, the number of divisions at each stage being constant for a given species.

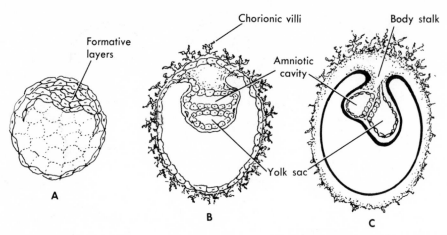

Fig. 5-2. Very early development of the embryo. **A,** Formative layers in the inner cell mass. **B,** Further differentiation of the formative layers and beginning formation of amniotic cavity. **C,** More differentiation of embryonic structures: formation of body stalk and localization of chorionic villi.

As cleavage continues, cell division becomes unequal in both size and configuration. Some of the cells thin out to form a hollow fluid-filled ball called a *blastocyst* while others form a cluster of cells, the *inner cell mass,* that protrudes into the cavity. Even at this very early stage there is evidence that the blastocyst is a highly organized structure with some cells dividing faster than others. From two *formative layers* within the inner cell mass the embryo proper will develop while the remainder of the cells surrounding the cavity go on to produce the external embryonic structures. Within the inner cell mass the early embryo is rapidly differentiating into two adjoining layers of cells known as the ectoderm (*ecto,* outer; *derm,* skin) and the entoderm (*ento,* inner). The ectoderm is destined to become the brain, nervous system, sense organs, and skin. The inner tissues, including the lining of the digestive tract, the liver, pancreas, bladder, and lungs, develop from the entoderm. Soon some primitive cells appear in the region between the ectoderm and entoderm. These cells, the mesoderm (*meso,* middle), will become the remaining body structures: bones, muscles, the heart, blood and blood vessels, and the connective tissues. These three layers of cells that differentiate into the various organs of the body are known collectively as the *primary germ layers.*

The blastocyst releases enzymes that digest a small cavity in the lining of the uterus into which it becomes *implanted* when about 9 days old. The implanted blastocyst grows rapidly as its outer layer develops fingerlike projections called *villi* that penetrate the uterine lining and absorb food and oxygen from the mother's blood (Fig. 5-2). This outer layer of cells differentiates into a membrane, the *chorion.* The largest portion of the chorion will become smooth as the contents within grow and expand, while the portion whose villi continue to penetrate the uterine lining will proliferate to form an important structure, the *placenta,* through which the developing organism will derive vital materials during intrauterine life.

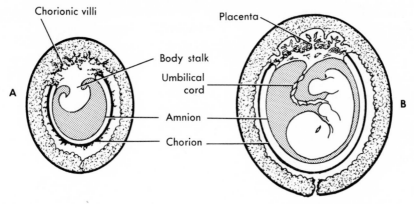

Fig. 5-3. The embryo, **A,** and fetus, **B,** in relation to fetal membranes and intrauterine structures.

Within this chorion two cavities are forming: an amniotic cavity forms above the formative layers of the embryo, and a yolk sac develops beneath it. Some of the ectoderm cells form a second surrounding membrane as the amniotic cavity expands and envelops the entire embryo, which remains attached to the chorion by a *body stalk.* This membrane is the *amnion* and contains the *amniotic fluid* (Fig. 5-2).

The three germ layers continue to grow and differentiate until at the end of 8 weeks the embryo looks quite human. The yolk sac has almost disappeared and the developing organism, floating in its amniotic fluid, is encased in two membrane layers: the amnion and the chorion. The embryo derives its nourishment through a body stalk connecting it to the chorionic villi that penetrate the maternal system. The embryo is now a *fetus.* The body stalk elongates to form the umbilical cord containing blood vessels that carry blood from the fetus to the specialized portion of the chorion, the placenta, and back to the fetus (Fig. 5-3).

Differentiation

All somatic cells in an individual have the same complement of genes as the zygote from which they have descended, yet any single cell differs measurably in form and function from the zygote and from other differentiated cells. Through successive mitotic divisions identical sets of chromosomes and their genes are equally distributed to each daughter cell. As cell division proceeds, these cells gradually deviate and assort into a variety of very different cell types; for example, muscle, nerve, bone, and gland cells. These diverse cells, although phenotypically different, are presumed to be genotypically identical; therefore, only a small portion of its genes are manifest in each cell. For example, all cells contain the genes for synthesizing hemoglobin, insulin, and serum albumen, but only in specialized cells are they active—only erythrocytes synthesize hemoglobin, only islet cells of the pancreas synthesize insulin, and only liver cells synthesize serum albumen. In all other cells these genes are repressed.

Differentiation is the process by which cells acquire specific and characteristic properties. This is accomplished by various mechanisms—division, shifts in activity, tissue movement, increase in size, increase in number, cellular death, and so forth—in a specified sequential order. Each step in the differentiation process depends upon successful completion of a previous step. Anything, such as a mutant gene or environmental agent, that interferes with one of these steps will cause an arrest in the development of that particular tissue or organ. Divergence from the normal course of development will result in maldevelopment of a part or, if at an early stage, a sequence of distortions causing more severe or multiple malformations. The younger the developing organism the more vulnerable it is to adverse influences.

Basic cell differentiation. During very early times it was thought that the sperm cells contained the human being in miniature with organs and parts fully formed and that it needed only the nourishment of the intrauterine environment to grow and develop. Later assumptions pictured differentiation as a process whereby determinants of each cell type (including the genes) were apportioned to the correct cell line at the appropriate time during development. However, current understanding gained through experiments with lower animals indicates that development is *epigenetic;* that is, it consists of successive formation of and addition of new parts through an orderly pattern of development and interaction.

Differentiation depends not upon differences in the genes among cells but upon variations in the *activity* of the genes and the specific properties of cellular proteins. Genes in the nuclei of the early blastomeres of a developing embryo are distributed equally, but the chemical environment of the cell is not. The cytoplasm is a heterogeneous mixture of substances, and the distribution of this material is neither uniform nor random. With succeeding cell divisions the cytoplasm is partitioned into a number of smaller cells, each with slightly different chemical properties. Thus it appears that each of these different environments activates a *different* set of genes, the active genes then generate new cytoplasmic environments, and the interaction of the *active* genes with the *new* environment results in selective interaction or inhibition of new groups of genes. In this manner these early cells are driven along a predestined path of differentiation, and cell function becomes progressively more restricted. Whereas early embryonal cells are vague in function, with random form and structure, later embryonal cells are less flexible in function, diverse in size and structure, and gradually demonstrate a rigid lack of adaptability. Specialization has begun. Specialized tissues form, and from these the organs and organ systems develop.[5]

Tissue interaction and movement. There is a reciprocal relationship between tissues. Growth in the embryo is not uniformly progressive; changes in form and pattern in various regions of the developing organism occur at different rates. Whereas early differentiation is related to changes in the intracellular environment, now the alterations in cellular form and composition result from differences in the rate of division, tissue movement, and interaction between tissues. During the early stages of random cell motility in the embryo there appear to be significant interactions between two or more clusters of cells, or primitive tissues, and these interactions seem essential to normal tissue differentiation. One type of tissue has the ability to stimulate a response in a neighboring tissue, which then proceeds to differentiate into appropriate specialization. This interactive process can be one-sided or mutually inductive and responsive. Maldevelopment occurs if the stimulation of *inductor tissue* either lacks the capacity to induce a response, does not come into sufficiently intimate contact with the responding tissue, or the responding tissue lacks the capacity to respond to the inductor tissue.[8, 9]

Another closely related mechanism that is essential to development involves morphogenic movement (*morpho,* form; *genesis,* to produce). A group of cells, or primitive tissue, begins in one region and then at a specified time migrates to another region. Whereas tissue interaction brings about *specific types* of cell differentiation to assure that each cell type is formed at the right time and in the right place, tissue migration *relocates* the cells of different fates to assure order and shape during development. Such congenital anomalies as spina bifida or cleft palate are believed to result from abnormal movements during a critical stage of development.[5]

Cellular degeneration. Some phases of differentiation require cellular death as a normal characteristic at a specific point in time and space. Such cells are programmed to die and are an essential feature in the morphogenesis of many organs. Cell death is necessary to the formation of central canals, or lumens, in many organs that would otherwise remain solid or for the union or detachment of organ parts. For example, cell death is essential to such things as separation of fingers and toes, eyelids, and formation of ducts and is timed to take place at precisely the right stage in cell life. The sequence of events leading to cellular degeneration is subject to the control of genes and can be altered by defective gene action during such critical stages of development. A mutant gene that inhibits or prevents normal degeneration may allow the survival of cells such as those between the toes or fingers in syndactyly (webbed fingers or toes); normal cells in this region would die. Persistence of a small tail in some humans is attributed to lack of cellular death during development, and embryonic cells that fail to die are responsible for the development of some tumors of early childhood (for example, neuroblastoma). Abnormal degeneration will result in tissue loss at sites and stages not associated with normal morphogenesis, and excessive normal cellular death will create an abnormal spread beyond the usual limits of degeneration.[8]

Extrachromosomal effects. Production of protein depends upon RNA (p. 20). The RNA used for protein synthesis during early division of the zygote is probably maternal RNA present in the cytoplasm of the ovum. During oogenesis an enormous amount of cellular material is synthesized and stored in the egg cytoplasm and is subsequently passed on to the fertilized ovum. The first cell divisions following fertilization take place at a very rapid rate. These cytoplasmic materials, including RNA, are used to sustain the embryo during this early development while coding activity in the zygotic nuclei remains dormant. In this way the genetic constitution of the mother affects early development of the embryo. As development progresses, the embryonic nuclei begin to synthesize RNA from their own DNA templates.

There is interest and research regarding the role of the ovular cytoplasm in relation to early embryonic development. If the cytoplasm of the egg is deficient in any essential enzyme or substrate needed for growth and development, it may have adverse effects upon the embryo. The existence of such a mechanism has been demonstrated in some plants and lower animals and may provide insight into the etiology of some congenital defects.

Regulation of gene action

The manner in which genetic information is translated into proteins is well known (p. 20). A large number of different enzymes and other proteins are required for growth and development; therefore, the activity of structural genes must be able to be modified to meet the changing needs of the embryo during differentiation. Less is known about this selective control of protein synthesis in relation to time and place, but there must be some mechanism designed to turn structural genes on or off according to the cellular needs of the moment. It is now universally accepted that the structural genes responsible for the synthesis of polypeptides are under the control of an operator-regulator mechanism. There appear to be *operator* genes that are responsible for initiating protein synthesis and *regulator* genes that *repress* or *derepress* the operator genes. The repressors act through the cytoplasm to repress the operator genes, which respond by turning the structural genes off and thus repress protein synthesis. When the activity of the repressor is inhibited, the operator is *derepressed* and is then free to initiate protein synthesis. By this mechanism gene activity is switched on and off at appropriate times in response to unknown cytoplasmic signals.

Defective gene action can alter the course of events at any point in the development and differentiation process. Alterations in amino acid sequence (mutations) are an important cause of variation in protein structure, and whether the change is interpreted as normal or abnormal depends on the contribution of the protein to the fitness of the organism. Since regulator and operator genes are also composed of DNA they are subject to mutation. Harmful mutations may produce defective development such as persistence of an inappropriate protein or failure to initiate production of an essential protein.

Changes in protein during development

Changes in cellular proteins reflect changes in the activity of specific genes; therefore, changes observed in protein composition of cells at successive stages of development are useful indicators of gene control of differentiation. The changes in lactic acid dehydrogenase and hemoglobin serve as illustrations of proteins that undergo changes during differentiation as a result of a system of genetic switch mechanisms.

Lactic acid dehydrogenase isozymes. LDH is an enzyme protein that is important in cellular respiration and exists in several molecular forms called *isozymes* (or *isoenzymes*). These different forms all consist of four subunits, or tetramers, each composed of two types of polypeptide chains, A and B, that are encoded by two different genes. Each isozyme is designated by numbers 1 through 5, and its composition is determined by the number of A and B chains that it contains, so that five different combinations are possible. For example, the isozyme LDH-1 contains 4 B chains, the isozyme LDH-3 contains 2 A chains and 2 B chains, and the isozyme LDH-5 contains 4 A chains. These isozymes are distinguishable by the rate at which they migrate during electrophoresis, a technique whereby molecules are separated according to their size and electric charge. The most rapidly migrating isozyme is LDH-1 and the slowest is LDH-5.

ISOZYME:	LDH-1	LDH-2	LDH-3	LDH-4	LDH-5
COMPOSITION:	B_4	A_1B_3	A_2B_2	A_3B_1	A_4

(+) Most rapidly migrating ⟵—————————————— Most slowly migrating (−)

All tissues contain LDH, but the amount of each type shows progressive changes during fetal and postnatal development. The presence of any one of these isozymes is characteristic of that tissue at a specific stage of development. Embryonic tissues and the adult heart muscle contain predominantly type B_4 (LDH-1); but as development proceeds, the ratio of A and B subunits gradually shifts until type A_4 (LDH-5) is more active in adult tissue, especially skeletal muscles.[5, 6]

Development of hemoglobin. Hemoglobin provides an example of one gene acting during fetal development and another in extrauterine life. Hemoglobin is synthesized in erythroblasts, the early precursors of erythrocytes, or red blood

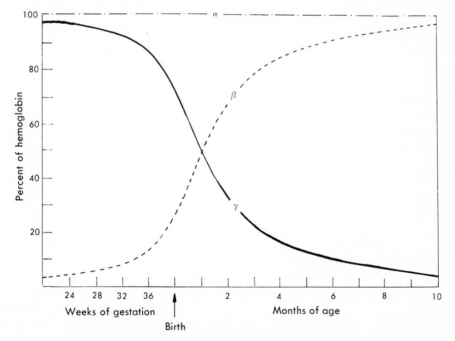

Fig. 5-4. Changes in percentage of fetal and adult hemoglobin. (From McKusick, V. A.: Human genetics, ed. 2, Englewood Cliffs, N. J., 1969, Prentice-Hall, Inc.)

cells. Erythroblasts synthesize two types of hemoglobin: hemoglobin A (HbA) and hemoglobin F (HbF). Each is encoded by different structural genes and differs in the amino acid sequence of its polypeptide chains, which is reflected in its physical, chemical, and functional characteristics. For example, fetal hemoglobin (HbF) has a greater capacity for transporting oxygen, which provides an advantage in the low oxygen environment of the intrauterine period.

Both hemoglobins are composed of four polypeptide chains: adult hemoglobin (HbA) contains two alpha and two beta chains, while fetal hemoglobin (HbF) consists of two alpha and two gamma chains. Synthesis of α-polypeptide chains begins in the embryo and remains functional throughout the life of the individual, but there is a reciprocal change in the ratio of γ and β chain synthesis during differentiation. Gamma chain production begins at the same time as α chain synthesis in the embryo, but the rate begins to decline in the latter part of fetal life until HbF disappears at about 4 months of age. Concurrently, production of β chains begins and, as γ chains synthesis declines, β chain synthesis increases until all hemoglobin is HbA (Fig. 5-4).

Twins

An abnormality of early development that occurs with relative frequency is twinning. Twins have been of interest since antiquity and have provided an appealing theme for dramatists and novelists (for example, Shakespeare's *Comedy of Errors* and Dumas' *Corsican Brothers*). The distinctive characteristics of the two types of twins have also been of special interest to both geneticists and environmentalists in their efforts to obtain information regarding the "nature-nurture" controversy.

Types of twins. It is well known that twins are of two distinct types: identical, or monozygotic (MZ), and fraternal, or dizygotic (DZ). These two types are separate and apparently unrelated phenomena. Dizygotic twins are derived from the fertilization of two ova that are released nearly simultaneously by the mother. They may be of like sex or opposite sexes, and they differ in both genotype and phenotype. Monozygotic twins are the result of one fertilized ovum that becomes separated at a very early stage of development, with each part developing into a complete individual. They are both genotypically and phenotypically alike—always the same sex. The term identical, although often used to describe monozygotic twins, is not entirely accurate, because no two individuals are ever exactly alike in every detail.

INCIDENCE. The frequency of twin births varies according to ethnic origin, maternal age, and heritability and is primarily due to differences in the incidence of dizygotic twins. Monozygotic twins occur with relatively uniform frequency in all populations and are considered to be random events; the incidence is approximately 1 in 200 to 285 births. Dizygotic twinning, on the other hand, shows variable frequency among racial populations, the highest being in the Negroid races (as high as 1 in 20 to 30 pregnancies in some African races), lowest in the Mongoloid races (approximately 1 in 150 to 200), with Caucasoids somewhere intermediate (1 in 80 to 85). In the United States the overall twinning rate is approximately 1 in 80 pregnancies and consists of one third monozygotic and two thirds dizygotic twins; in Japan this ratio is reversed.[1, 2]

Dizygotic twinning becomes increasingly common with advancing maternal age, rising to a maximum between the ages of 35 to 39 then decreasing rapidly. Maternal age has little if any effect on the monozygotic twinning rate. There is some evidence, although not conclusive, that there is a positive correlation between undernutrition and a decline in the rate of dizygotic twinning.[2]

HEREDITY. Monozygotic twinning is unaffected by heredity, but dizygous twins show a marked familial tendency. The chances of a woman who has had one pair of monozygotic twins producing a second pair are no more than they would be had she not had the first pair; however, women who have had one pair of dizygotic twins are more likely to have a second as compared with the general population. In other words, large families in which there were monozygotic twins show no increase in twin births, but families with dizygotic twins show a decided increase in the incidence of twins. Most studies of twins seem to indicate that the propensity for dizygotic twinning is confined to the female line. There appears to be an increase in twins among relatives of mothers of twins (for example, female sibs and offspring of dizygotic twins) but not among the relatives of the fathers (for example, brothers of dizygotic twins and offspring of a dizygotic twin). It is the genotype of the mother that affects the frequency of dizygotic twins among her offspring; the genotype of the father has no effect. Apparently the tendency toward double ovulation is an inherited trait expressed only in females. Fathers do, however, appear to transmit the disposition toward double ovulation to their daughters.

Double ovulation, or the simultaneous release of two eggs from the ovary, seems to be related to an increased production of pituitary gonadotropin in some races and in families where twinning is a familial trait. (Pituitary gonadotropin is necessary for maturation or ripening of germ cells in the gonads.) Studies of both animals and humans indicate that the number of ova released is controlled by the amount of pituitary gonadotropin and seems to be directly related to the size of the pituitary gland—the larger the pituitary, the greater the amount of gonadotropic hormone secreted. Observations that correlate with the incidence of dizygotic twinning and pituitary size are: the pituitary gland is larger in African races, reaches its maximum weight in the fourth decade, and increases in size with successive pregnancies. Also, the increased incidence of multiple births following administration of gonadotropin for infertility tends to bear this out.[1]

Determination of zygosity. It is important to distinguish between monozygotic and dizygotic twins for two reasons. First, monozygotic twin studies serve as a useful tool in the scientific study of the influence of heredity and environment in development phenomena and disease processes. Second, since there is an ever increasing need for transplant donors and truly successful organ or tissue transplantation is possible only between genetically identical individuals, identification of monozygotic twins is a very practical consideration. The earlier this distinction is made, the more useful the information will be. Methods used to determine zygosity are examination of fetal membranes or comparison and contrasting of physical similarities and differences between members of a pair. It can be established that a pair is not monozygotic, but not with absolute certainty.

Twins of different sex or with obvious differences in physical characteristics such as hair or eye color or ear shape are dizygotic. Monozygotic twins are always of like sex.* There may be observable differences in monozygotic twins, such as mirror-imaging; that is, one twin who is right-handed and one who is left-handed, or other minor asymmetries. (This phenomenon was believed to result if division took place relatively late, although this is not accepted by all authorities.) Blood group comparisons are the most reliable physical means to distinguish types of twins. Monozygotic twins always possess identical blood groups; dizygotic twins may be alike or they may differ in any or all blood group systems. If a single difference is found it can be concluded that the pair is dizygotic.

---

*Monozygotic twins of phenotypically different sexes have occurred due to loss of a Y chromosome during early division in one member of an XY twin pair. This produced one twin with an XY genotype and one with an XO genotype who had a female phenotype.

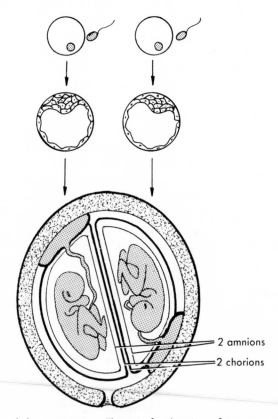

Fig. 5-5. Formation of dizygotic twins. There is fertilization of two ova, two implantations, two placentas, two chorions, and two amnions.

Examination of fetal membranes provides an early means of differentiating between monozygotic and dizygotic twins of like sex. Dizygotic twins have two separate and distinct placentas and membranes, both amnion and chorion (Fig. 5-5). In some instances, if the implantations are close together on the uterine wall the placentas may grow together, giving the impression of one placenta.

Monozygotic twins may have single or separate placentas and membranes, depending upon the time during early development when division has taken place. If, during the blastomere stage, the cells do not separate and two inner cell masses form, the two embryos will develop within a single chorion but with individual amnions. Rarely, the embryos develop within a single amnion. If the blastomeres separate, the two zygotes formed from this separation will implant separately and form their own amnion and chorion in much the same manner as dizygotic twins (Fig. 5-6). When division is late and incomplete the result is conjoined, or "Siamese," twins. Twins that are enclosed in a single chorion (monochorionic) can be regarded as monozygotic twins; however, in other cases distinction is not certain since both types of twins can have two amnions, chorions, and placentas or a single placenta.

Another method sometimes used to establish zygosity of like-sexed twins is skin transplantation. Reciprocal skin grafts are normally accepted in monozygotic pairs and almost universally rejected in dizygotic pairs.

Importance of twins in genetics. Twins have been used by scientists for years in their attempts to determine the roles played by heredity and environment in the development of specific traits. The twin method has been of greatest value in providing information concerning genetic predisposition to the development of a disorder. A disease or trait appearing with a higher frequency in monozygotic than in dizygotic twins suggests a genetic etiology. Since both monozygotic and dizygotic twins normally grow up under the same or similar conditions, any differences that appear in dizygotic but not in monozygotic pairs can be attributed to differences in genotypes. On the other hand, differences appearing in both monozygotic and dizygotic twins must be the result of environmental influences although these are usually difficult to determine. If both members of a twin pair display the trait under study, they are said to be *concordant* for that trait; if only one of a twin pair displays the trait, they are said to be *discordant* for that trait. Twin comparisons are of greatest value in assessing characteristics or traits that are not due to a single gene but are felt to be the result of several genes or factors exerting an effect on the phenotype, such as some congenital malformations and infectious diseases. They have also been useful for testing some therapeutic procedures, with one monozygotic twin serving as experimental and the other as the control.

Multiple births. Twins are not uncommon in the population, but triplets are rare and quadruplets or quintuplets are extremely unusual, as evidenced by their news appeal. The incidence of these multiple births is estimated according to *Hellin's law;* that is, if twins occur once in 87 births, then the frequency of triplets will be one in 87^2 births and quadruplets one in 87^3 and so on. These higher multiple births can originate in a variety of ways: from a single ovum, from separate ova, or from a combination of the two types (Fig. 5-7).

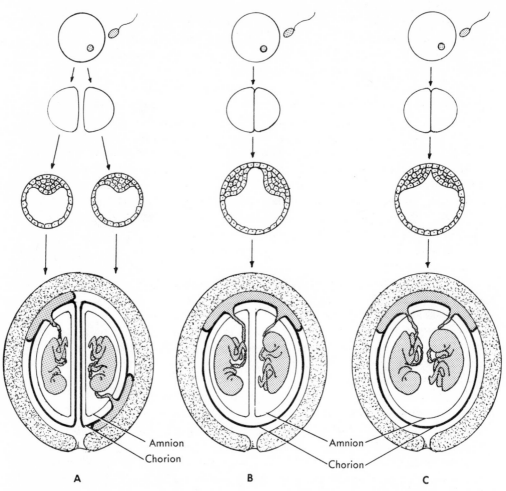

Fig. 5-6. Formation of monozygotic twins. **A,** One fertilization, blastomeres separate, resulting in two implantations, two placentas, and two sets of membranes. **B,** One blastomere with two inner cell masses, one fused placenta, one chorion, separate amnions. **C,** Later separation of inner cell masses, with fused placenta and single amnion and chorion.

Aging

Very little is known about the relationship between genetic endowment and the aging process. Although the length of life of an organism depends primarily on its heredity, there is no evidence that there are specific genes for aging or that there are single genes that can lengthen life. It is known that the life span of an organism is species-specific and that physical changes occur at different rates for different organs, remaining constant for that organ from individual to individual. Individuals in some kindreds have a significantly longer life span than others, and

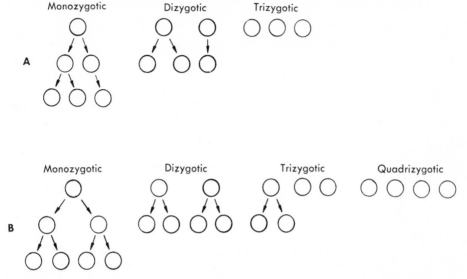

Fig. 5-7. Formation of triplets and quadruplets, indicating the variety of mechanisms that can produce multiple births. Triplets can be formed from one, two, or three ova, **A**; quadruplets from one to four ova, **B**.

there is a smaller difference between the life span in monozygotic twins than in dizygotic twins of like sex.

Numerous theories have been advanced regarding the causative factors in aging. Aging may be simply the result of cellular "wear and tear" or due to progressive mechanical-chemical effects on cellular function. The increasing accumulation of somatic mutations and loss of DNA as well as progressively increasing auto-intoxication (destruction of tissue as the result of uneliminated toxins generated by the body) have also been suggested as causes. All indicate that, with age, an organism is increasingly unable to adapt to the environment and to maintain homeostasis. Aging might be viewed as an integral part of the developmental continuum—a decline in growth that represents one end of the spectrum of differentiation.

Cellular death accounts for functional loss in some tissues, but alterations in function of surviving cells is difficult to assess. Of interest to geneticists is the detection of chromosome loss in cells of older persons. It has been found that older persons, male and female, demonstrate an increased incidence of cells with 45 chromosomes. The chromosome losses are predominantly the X chromosome in females and the Y chromosome in males. Based on knowledge of age-related non-disjunction in disorders due to chromosomal aberrations (Chapter 4) this is not surprising. Loss of an X or a Y chromosome is not a threat to life; however, the question arises whether or not these events might reflect loss of more and larger chromosomes and subsequent cell death in all tissues.

SEX DIFFERENTIATION AND DEVELOPMENT

Genetic sex is determined at conception and depends upon whether the ovum is fertilized by a sperm bearing an X chromosome (female phenotype) or a Y chromosome (male phenotype).

Sex ratio

The sex ratio is the ratio of the number of male births to the number of female births. Since sex is determined by whether the ovum is fertilized by an X-bearing or a Y-bearing sperm, it would be expected that male and female offspring would be produced in equal numbers. However, the *secondary sex ratio* (the ratio of sexes at birth) is 106 males to every 100 females. The *primary sex ratio* (the ratio of the sexes at fertilization) is estimated to be much higher at approximately 130 to 100, as evidenced by sex chromatin studies of early embryos. The reasons for this prenatal selection against males are unknown, but it has been suggested that it might be due to X-linked recessive lethal chromosomes or that the two X chromosomes provide the female with heterotic (hybrid) vigor. This female advantage persists throughout the life span. The higher male mortality continues until the male-female ratio becomes 1 to 1 at the age of reproduction during the third and fourth decades. Then, the excess of females becomes more and more pronounced until by the age of 85 the ratio becomes approximately 85 males to every 100 females.

There are families recorded in which there is a preponderance of one sex, and in most instances this is probably no more than chance. However, there are rare families in which the offspring for numerous generations are of one sex exclusively. In a strictly male pedigree it is assumed that the only functional sperm produced by the males are Y-bearing sperm. In exclusively female pedigrees the females apparently transmit to their daughters some property that allows fertilization of their eggs by X-bearing sperm only.

Sex differentiation

Sex differentiation refers to the development of gonads, internal ductal system, and external genitalia. For the first 6 weeks of gestation the developing embryo is morphologically neutral, neither male nor female. It possesses primitive gonads consisting of an outer layer, the cortex, and an inner medulla; two pairs of ducts, the Mullerian and Wolffian ducts; and a bipotential region that will be transformed into the urogenital structures.

In genetic males, during the seventh and eighth weeks of intrauterine life the gonadal *medulla* develops and the cortex regresses, the Wolffian duct system forms the duct systems of the testes while the Mullerian system remains rudimentary, and the primitive genital structures begin to be transformed into the male external genitalia. In genetic females the gonadal *cortex* develops as the medulla regresses; the Mullerian system proliferates to form the oviducts, uterus, and vagina while the potentially male ducts are arrested; and the female genitalia begin to take shape. The development of male characteristics is apparently determined by genes on the Y chromosome. In the absence of either ovarian or testicular stimulation there is a tendency for the embryo to differentiate into a female. The male genitalia do not develop without a functional testis, but it is unclear whether this is due to the Y chromosome alone or in collaboration with as yet undetermined influences.

Aberrant sexual development

Anomalies resulting from disturbances in sexual development can be classified as chromosomal and nonchromosomal disorders. Anomalies due to sex chromosomal aberrations were considered in the previous chapter and are primarily due to a departure from the normal number of sex chromosomes. Sex abnormalities not related to the sex chromosomes may arise from aberrant development at any stage in the differentiation process as a result of hormonal or other environmental influences.

Most problems of sexual ambiguity are encountered at birth or in early childhood. It is of utmost importance that the correct sex be established as early as possible in order that corrective procedures can be instituted soon enough to provide the individual with an appropriate sexual identity. It is the opinion of many authorities that the gender role is firmly established by 18 months of age. Although early assignment is the goal, sometimes the necessary diagnostic procedures may delay a final determination for several days or weeks.

Individuals whose sex is ambiguous or doubtful are often termed *intersex,* broadly defined as one whose phenotypic sex differs from the true genetic sex. An intersex fails to display all the criteria normally associated with a member of his or her sex. The major classifications of intersex in humans are *male pseudohermaphrodites, female pseudohermaphrodites,* and *true hermaphrodites* (Gr. Hermaphroditos: Herm[es] + Aphrodite, god and goddess of Greek mythology).

Hermaphroditism. True hermaphrodites are rare. They may be either genetic males or genetic females but have *both* testicular and ovarian tissues that can be recognized by histologic examination. The internal and external sexual organs are variable although the majority of hermaphrodites are chromatin positive with a 46,XX chromosome constitution. Sex chromatin tests and hormone studies are of little value in diagnosis. The diagnosis is made when biopsy reveals both male and female gonadal tissue in the same individual. The cause is unknown.

Male pseudohermaphroditism. Male pseudohermaphrodites are true genetic males (XY) with testes, usually inguinal, but whose external genitalia and secondary sex characteristics are to varying degrees female. A rare but well-understood anomaly of male sexual differentiation is testicular feminization.

Testicular feminization syndrome. Individuals with testicular feminization are males who are female in external appearance. They are almost always reared as girls and are usually not recognized as genetic males until they seek advice for amenorrhea. These individuals are frequently very attractive, develop breasts, and have a small vagina that ends in a blind pouch. Very often they are married. The testes are present either abdominally or in the inguinal canal and often diagnosed as hernias, and their chromosome constitution is XY. It is uncertain whether the testes secrete estrogen rather than testosterone or whether there is a defective response at the peripheral cellular level (including fetal life) that prevents male differentiation. Testicular feminization is definitely hereditary, but some doubt exists as to whether the inheritance pattern is X-linked recessive or sex-limited autosomal-dominant since affected persons do not reproduce. Removal of the gonads is recommended after puberty (due to a high incidence of gonadal tumor formation) and subsequent administration of estrogens. The psychosocial orientation is always female and should remain so.

Female pseudohermaphroditism. The most frequent cause of female pseudohermaphroditism is related to increased secretion of androgens from the adrenal cortex. These individuals are true genetic females (XX) with varying degrees of male phenotypic differentiation. The male manifestations may be mild or so severe as to produce external genitalia that are indistinguishable from the cryptorchid male. Overdevelopment of the adrenal cortex due to an inherited defect in steroid metabolism was discussed in Chapter 3. The external genitalia of the female fetus may also be altered by male or female sex hormones that reach the fetus through the maternal circulation. The source of these hormones can be overproduction by the maternal adrenal cortex or from hormones administered to the mother therapeutically. For example, some progestational agents, frequently given in an effort to prevent spontaneous abortion, are known to have androgenic properties.

CONGENITAL MALFORMATIONS

Congenital malformations, sometimes called *birth defects,* are defined as structural defects present at birth. The incidence is estimated to be about 6 in every 100 births, but the methods of interpretation and reporting are subject to the criteria of the investigator or recorder. Congenital defects have been noticed and described for thousands of years as evidenced by primitive drawings and carving and descriptions on ancient clay tablets. The striking appearance of abnormal human development has served as the origin of numerous legendary and mythologic creatures such as dwarfs, Cyclopes, double-headed deities, and so forth.

The development of an organism, especially during embryogenesis, is an intricate process. Growth and development of all parts must be properly integrated to ensure a coordinated whole. The rate must be such that one part is ready when needed by another part, otherwise either part may cease to grow or may deviate from its normal path. For example, during early development the optic lens is formed from overlying tissues only after an inductive trigger is released by the optic vessel. Some malformations result when a state, present in one phase of development as a normal condition, persists into another phase as abnormal. For example, a harelip is normal in a young embryo and a patent ductus arteriosus is essential during prenatal life. Any agent that interferes with these complex processes will produce a defect in development ranging in severity from complete degeneration to a local anomaly. Malformations can arise at any stage of development and present wide variability in the determining factors as well as the type, severity, and frequency of defects.

Congenital malformations can be broadly classified according to their probable etiology: (1) those that are determined by a single mutant gene on either an autosome or a sex chromosome, such as polydactyly or deafness (Chapter 3); (2)

those caused by chromosomal aberrations, such as Down's syndrome (Chapter 4); (3) those due to intrauterine environmental factors, such as heart defects as a result of maternal rubella or absence of a limb from maternal ingestion of the drug thalidomide; and (4) those that are the result of a complex interaction between genetic and environmental factors, such as clubfoot, or dislocated hip (Chapter 6).

Environmental influences on development

The growth and development of an individual is the result of a continuous interplay of genetic potentialities and environmental influences. No matter what the genetic endowment of the individual, an unfavorable environment may result in underdeveloped genetic potential. For example, a man who received genes for tall stature but who had inadequate nutrition and stimulation may evidence only a portion of his genetic potential; on the other hand, a man who inherited genes for short stature will be short no matter how superior his environment. Mutant genes frequently exert their effects only when environmental factors come into play. For example, the deleterious effects of galactosemia are manifest only after the individual with defective genes consumes milk sugar, and the gene for sickling has no ill effects upon the heterozygous carrier until he encounters an atmosphere of low oxygen concentration. Persons who inherit a tendency to allergic conditions will be unaffected until they encounter the environmental conditions that produce a physical response.

Genetic influences exert their effects primarily during the early stages of development when the various tissues, organs, and systems are becoming established. Environmental factors assume more importance as growth and development proceed throughout the life span. During intrauterine life the developing organism is protected to a great extent by the environment provided by the mother; however, this protection is not complete. Numerous internal and external factors can produce injury to the embryo especially during periods of rapid growth or differentiation. The impact of these factors depends upon the nature of the environmental change and the developmental stage of the organism at the time of exposure. The stages of accelerated growth and differentiation during which specific organs and systems are particularly vulnerable to environmental insults have been termed *critical periods,* and the agents capable of producing an adverse effect are called *teratogens* (*terato,* monster; *genesis,* production). Susceptibility to environmental influences decreases as organ formation advances. The younger the organism and the fewer the number of cells, the greater the extent of involvement proportionately when any cell or group of cells is injured (Fig. 5-8).

Teratogenesis. This term refers to the origin or method by which growth processes are disturbed to produce a physical defect; *teratology,* a comparatively new discipline, is the study of defective development. From the time the ovum is fertilized by the sperm until approximately 3 months' gestation is the period during which all the major organs and systems are laid down and begin to function. The remainder of the gestation period consists of refinement and development of these tissues. The approximate times of critical differentiation for some of the major or-

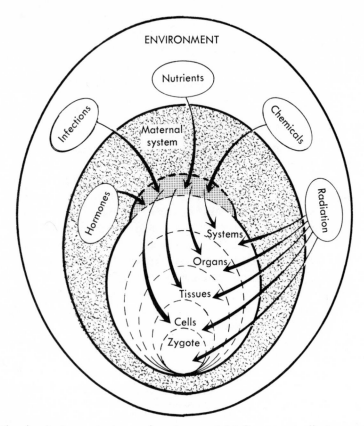

Fig. 5-8. The developing organism and environmental influences at all stages of development.

gans and systems have been identified (Fig. 5-9) from data gathered by retrospective studies of congenital abnormalities and animal experiments. All of the major organ systems begin development shortly after the blastomere stage, approximately the time of implantation in the uterine wall. The pregnant woman has not yet missed her first menstrual period!

Teratogens may act directly on the embryo or its accessory structures (for example, the placenta) or indirectly by its effect on the maternal system. Most teratogens produce their most profound effects during the limited period of rapid organogenesis. Since they must affect a specific process in the developing embryo, the *time of application* will determine the type and extent of the damage. For example, major damage from the rubella virus appears to occur in the first 4 weeks of pregnancy. The incidence falls during the third month and is relatively harmless to the fetus for the remainder of gestation. If the infection occurs in the sixth week, cataracts may develop; if in the ninth week, the child may be deaf; and heart defects can result from infection between the fifth and ninth weeks. Thus, a single teratogen may produce a variety of defects. Also, a variety of teratogens

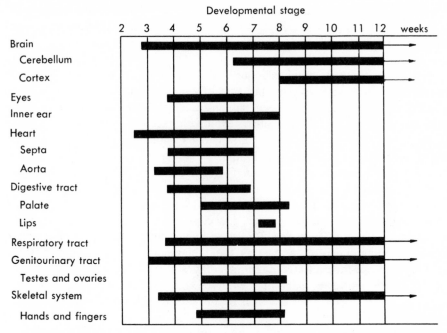

Fig. 5-9. Approximate periods of critical differentiation for some specific organs.

may produce similar anomalies (for example, viruses, chemicals, and x-rays, can all produce a mental deficit). Another factor that appears to play an important role, as evidenced by animal studies, is the genotype of the organism. Some species and strains of the same species have significantly greater susceptibility to a given teratogen.

Many teratogenic agents have little or no adverse effect on the maternal organism and may even be beneficial to the mother. The dramatic illustration is represented by the well-publicized defects produced by thalidomide: this drug, nontoxic to the mother, is severely teratogenic to the fetus. The best-known teratogens in man are infectious agents, radiation, and chemicals.

Infectious agents. The best known of the infectious agents proved to cause birth defects in man is the *rubella,* or German measles, virus. This important discovery was made in the early 1940's when it was found that the increased incidence of some congenital malformations correlated with an epidemic of German measles at about the time the mothers of these babies were in their third month of pregnancy. The most common defects found in these infants are eye defects (especially cataract), heart defects, deafness, microcephaly (small size of head and brain), and mental retardation. As mentioned previously, the type of defect is directly related to the stage of fetal development at the time of exposure. The incidence and degree of malformation in affected infants is unrelated to the severity of illness in the mothers: a mother who is mildly affected may bear a child with severe and extensive defects; in contrast, slight defects may follow a severe attack of maternal rubella. The consistency with which the virus produces malformations and the seriousness of such defects are sufficient reason to recommend termination of a pregnancy when a mother contracts rubella during the critical period of development.[7]

Numerous other maternal infections, such as the viral diseases of mumps, influenza, chicken pox, smallpox, and hepatitis, and a disease due to a protozoan parasite, toxoplasmosis; have been demonstrated to interfere with fetal development. Babies born to syphilitic mothers exhibit characteristic defects.

Radiation. Radiation has been shown to be both mutagenic (capable of producing mutations) and teratogenic in man. Pelvic irradiation of pregnant women—through natural background radiation (present everywhere in varying degrees), in those occupationally involved in its use, and those being treated diagnostically or therapeutically—has been proved to be hazardous to the embryo. As with other teratogens, the type of effect produced is closely correlated with the stage of development at which it is applied. Radiation may damage the organism at any time during its prenatal existence. During the preimplantation period, when there is rapid division but little differentiation, irradiation produces a high incidence of embryonic death; however, surviving embryos are normal. Radiation during the period of major organogenesis, which begins shortly after implantation, can cause a whole spectrum of malformations depending on the exact stage of development at which the radiation is applied. Most of these malformed embryos survive.

After the major organ systems have been formed, irradiation can still cause injury to the organism. The malformations are less severe than those brought about during the earlier periods, and the doses required to produce them are usually higher. The cells of the nervous system seem to be most sensitive at this time. Defects are primarily related to specific neural structures within the system; for example, the cerebellum and retina rather than the brain or the system as a whole. The verification that radiation is damaging to the embryo and fetus is sufficient to contraindicate the use of x-ray during pregnancy and in young women during the period in the menstrual cycle that implantation is likely to occur.

Chemicals. The thalidomide tragedy of the early 1960's has accelerated the interest in environmental factors as a cause of congenital defects. Many drugs have been suspected to have teratogenic properties, but none have created the impact of thalidomide. During the few short years that it was available it is estimated that four to five thousand children were affected in Germany alone (the country in which it was most widely used). Although an efficient sleep-inducing drug in the mother, it produced such characteristic malformations in the fetus as phocomelia (seal-like extremities) or amelia (absence of a limb), primarily of the arms but involving all extremities; some facial defects; heart malformations, especially atrial or ventricular septal defects; and visceral abnormalities such as absence of the gallbladder and double uterus. When the drug was taken during the sensitive period, between the thirty-fourth and fiftieth days of gestation, the incidence of malformation reached 100% even with small doses. As soon as the relationship between thalidomide and defective babies was recognized, the drug was immediately withdrawn from the market.

The list of drugs implicated in birth defects is increasing in length although not all have been proved beyond a doubt to cause malformations. Cancer chemotherapeutic agents, because they are effective in destroying the rapidly dividing cancer cells, can exert the same influence on the rapidly dividing cells of the embryo, and chromosome damage has been demonstrated to occur in individuals who are taking some types of drugs. The one that has received the most attention is lysergic acid diethylamide (LSD), the hallucinogenic drug used by many young people of child-bearing age. LSD has been demonstrated to produce chromosome breaks in laboratory animals, and such breaks have been observed in therapeutically aborted fetuses in a small group of LSD users. However, there have been so few studies conducted in humans that it is still controversial whether LSD ingested during pregnancy is capable of inducing abnormalities in growth and development.[4]

Nutritional factors. Disturbances in nutrition can influence *all* stages of growth and development. There is substantial documentation regarding the relationship between the diet of the pregnant woman and the physical condition of her newborn. During the prenatal period, from implantation to birth, interruption of nutrition to the embryo can result from maternal deficiencies or from defective extra-uterine structures. Faulty implantation of the ovum, degeneration of the chorion, or an alteration in location or function of the placenta can interrupt the nutrition of the developing organism. Any of these can result in abortion, malformation, or retarded growth. Maternal dietary deficiencies, either total reduction of food intake or lack of specific nutrients, is sufficient to cause fetal damage. Early development involves an increase in cell number followed by an increase in cell size; therefore, a deficiency in essential nutrients during the earlier stages of development will produce a permanent deficit in overall cell numbers. Undernutrition during the period of rapid cell proliferation and differentiation is felt to cause significant central nervous system deficit.[10]

Phenocopies

There are many congenital defects due to nongenetic factors. A phenocopy is a phenotype produced by environmental factors that imitates or is indistinguishable from one genetically determined. For example, deafness, cretinism, or congenital cataract can all be due to mutant genes but they may also be caused by exogenous agents. Deafness can be a result of a number of teratogens, rubella virus can cause congenital cataracts, and lack of iodine in a child can produce cretinism. Mental retardation presents a particularly difficult problem in assigning a cause. It is a manifestation of a variety of syndromes, both single-gene and chromosomal, and numerous environmental agents are known to be damaging to brain tissue; for example, lack of oxygen as a result of anesthesia or drugs during labor and delivery. An environmental agent can mimic almost any genetic phenotype. For this reason it is extremely important that such exogenous factors are ruled out before any given congenital defect is labeled as hereditary.

Abortion

Although the actual incidence of spontaneous abortion is in doubt and estimates vary considerably, probably 10% to 25% of human pregnancies are lost within the first 3 months following fertilization. In most instances the causes are never determined to any degree of satisfaction. There is agreement among authorities that loss of a normal embryo is most likely the result of an unsuitable maternal environment, but it is impossible to calculate the number of these and the number of abnormal zygotes that never reach the implantation stage.

Lethal genes (those that produce their effects at an early stage of development) probably account for a significant number of abortions although most of the evidence for this has been provided through animal studies. Exogenous agents such as mechanical factors, toxic chemicals, and endocrine imbalance, especially those produced by the placenta, have been implicated in abortion. There has also been some evidence of an increase in the abortion rate during epidemics of some infectious diseases; for example, rubella. Ionizing radiation can cause early death depending on the dosage and the age of the embryo.

There is evidence to indicate that antigen-antibody incompatibility is a cause of intrauterine death. It is known that mothers with type O blood have natural antibodies against A and B antigens of fetal tissues, and the increased loss in many ABO incompatible matings probably represents an antigenic reaction to the foreign tissues—the mother literally rejects the fetus. Rh incompatibility causes death at a later stage in development but does not seem to be a factor in abortion (p. 122).

The most common identifiable cause of early fetal loss is chromosomal aberrations (see Chapter 4). Chromosome disorders are felt to be the most frequent cause of all congenital disorders in man, but fortunately about 90% of embryos so affected are lost through spontaneous abortion. In the investigation of early abortuses, chromosome abnormalities were found in 20% to 40%. Trisomy was the defect most frequently encountered in abortuses (approximately one half) usually involving the D, E, and G group autosomes and trisomies of some chromosomes that are extremely rare in live-born infants were found in significant numbers. Multiple chromosome abnormalities have been seen in fetuses who reach a size that can be termed an abortus; other embryos are presumably lost before abnormalities can be recognized. Absence of a sex chromosome (the 45,X or XO genotype) accounted for a large number, indicating that the XO genotype is lethal. This is the least frequently occurring sex chromosome abnormality found in live infants. There were a significant number of abortions at the extremes of maternal age and among translocation carriers.[3]

REFERENCES

1. Benirschke, K.: Origin and clinical significance of twinning, Clin. Obstet. Gynecol. **15:**220, 1972.
2. Bulmer, M. G.: The biology of twinning in man, Oxford, 1970, Clarendon Press.
3. Carr, D. H.: Chromosomes and abortion. In Harris, H., and Hirshhorn, K.: Advances in human genetics 2, New York, 1971, Plenum Publishing Corp.
4. Dishotsky, N. L., and others: LSD and genetic damage, Science **172:**431, 1971.
5. Markert, C. L., and Ursprung, H.: Developmental genetics, Englewood Cliffs, N. J., 1971, Prentice-Hall, Inc.
6. McKusick, V. A.: Human genetics, ed. 3, Englewood Cliffs, N. J., 1969, Prentice-Hall, Inc.
7. Rhodes, A. J.: Effects of viruses. In Fishbein, M., editor: Birth defects, Philadelphia, 1963, J. B. Lippincott Co.
8. Saxen, L., and Rapola, J.: Congenital defects, Philadelphia, 1969, Holt, Rinehart and Winston, Inc., Chapter 4.

9. Timeras, P. S.: Developmental physiology and aging, New York, 1972, Macmillan Inc.

10. Winik, M., Brasel, J., and Velasco, E. G.: Effects of prenatal nutrition upon pregnancy risk, Clin. Obstet. Gynecol. **16:** 184, 1973.

GENERAL REFERENCES

Emery, A. E. H.: Heredity, disease and man, Berkeley, 1968, University of California Press.

Hsia, D. Y-Y.: Human developmental genetics, Chicago, 1968, Year Book Medical Publishers, Inc.

Neurath, P., and associates: Chromosome loss vs. chromosome size, age and sex of subjects, Nature **225:**281, 1970.

Ohno, S.: Maternal influence upon genetic activity of early embryos. Birth defects, original article series **4**(6)**:**45, 1968.

Stern, C.: Principles of human genetics, ed. 3, San Francisco, 1972, W. H. Freeman and Co., Publishers.

Summitt, R. L.: Differential diagnosis of genital ambiguity in the newborn, Clin. Obstet. Gynecol. **15:**112, 1972.

Thiede, H. A.: Cytogenetics and abortion, Med. Clin. North Am. **53:**773, 1969.

Thompson, J. S., and Thompson, M. W.: Genetics in medicine, Philadelphia, 1966, W. B. Saunders Co.

Waranky, J.: Congenital malformations, Chicago, 1971, Year Book Medical Publishers, Inc.

6

Genes and immunity

Immunogenetics is a relatively new branch of the increasingly diversified field of genetics and is concerned with the genetics of immunity; that is, antigens, antibodies, and their reactions. Previously, immunology was concerned primarily with problems related to resistance to diseases caused by infectious agents, but its importance to other areas of biology and medicine has been recognized and its principles are being applied to problems in other areas, the most dramatic being organ transplantation. The major areas that will be considered are those in relation to the blood group systems, histocompatibility, and autoimmune disease. First, a very brief overview of some basic concepts of immunity will aid in understanding the significance of genetic endowment to some important body responses.

CONCEPTS OF IMMUNITY

The human organism is continually in contact with environmental agents capable of entering the body and causing a major disturbance; therefore, it must have some means of combating these foreign elements. Immunity (L. *immunis,* safe) implies protection from certain risks. The term is usually applied to protection from disease and particularly to those mechanisms that provide protection from a specific infection. There are some naturally occurring substances in the body that act against invading agents and inactivate them before they can produce any damage to the organism. This natural resistance to infection is called *natural immunity*. Since he is not resistant to all possible alien substances, man has a mechanism, the *immune process,* whereby his body can develop an *acquired immunity* to a given substance. The process is characterized by the entry into the body of an *antigen* that stimulates the development of protective proteins called *antibodies,* which will then react *specifically* with that antigen whenever it enters the body.

114

Antigens and antibodies

An antigen is a substance, usually a protein, that is capable of producing an immune response. For a substance to be antigenic it must be foreign to the individual producing the antibody. Antigens are almost always large molecules—proteins or very large polysaccharides. Other smaller compounds, *haptens,* may elicit an immune response by combining with naturally occurring proteins in the body or other proteins before entering the body. Such substances are chemicals in dust, drugs, industrial chemicals, and so forth. Antigens on viruses and bacteria are of concern in prevention of infectious disease, but those that are important in genetics are the genetically determined antigens on the red blood cells and tissue cells of the body. The cells of man contain many antigens, each of which is determined by one or more genes.

Antibodies are serum proteins formed by cells of the lymphoid tissue as a result of antigenic stimulation, reacting specifically with that antigen. The antibodies are almost all gamma globulins and do not differ in fundamental structure from the naturally occurring gamma globulins responsible for natural immunity. There are thousands of antibodies, known as *immunoglobulins* (Ig). The five main classes that have been studied most extensively in man are: (1) immunoglobulin G (IgG), which comprises 80% of the immune globulins in normal sera and is valuable in neutralizing microbial toxins; (2) immunoglobulin M (IgM), a larger globulin, usually the first to respond to serious infections or following vaccinations, which is rapidly metabolized with the formation of IgG; (3) immunoglobulin A (IgA), a smaller globulin that is able to cross cell barriers and is abundant in secretions such as tears, saliva, and mucus of the intestinal tract; (4) immunoglobulin D (IgD), with an as yet unknown specific function; and (5) immunoglobulin E (IgE), which is consistently elevated in allergic responses such as rashes, asthma, and so forth.

The antigenic response

To combat the toxic or invasive action of antigens several reactions may take place between antigens and antibodies.

1. *Neutralization* (antitoxic reaction) occurs when the antigen is a toxin. The antibody reacts with the antigen to render it ineffective so it can then be disposed of by other cells.
2. *Precipitation,* one of the simplest reactions, occurs when soluble protein substances, through combination with the antibody, no longer remain soluble but separate out from the solution as a solid.
3. *Agglutination* is the clumping together and subsequent immobilization of insoluble substances, such as bacteria or red blood cells, when antibodies attach themselves to the surfaces of two or more antigens.
4. *Lysis* (or dissolving) of antigenic cells occurs when several antibodies attach to the cell surface and, through the mediation of an accessory substance (complement), cause the cell to dissolve and spill its contents into the body fluid.
5. If the combined efforts of antibody and complement do not cause direct

destruction of the invading cell, they serve to make the cell more susceptible to the destructive action of specialized cells, the *phagocytes*. This is called *opsinization*.

The *primary immune response* to antigenic stimuli takes place following the initial entry of an antigen. First there is a lag, or latent period, of 2 days to a week or longer during which no antibody can be detected in the circulation. At the end of this latent period there is a rapid increase in antibody in the circulation, and the level remains elevated for variable periods of time, after which it then declines. Stimulation with the same antigen at a later time (months or years after the initial contact) precipitates a *secondary response*. This response is characterized by a shorter latent period with a more rapid rise in and higher level of antibody than took place in the primary response, and the antibodies persist for a longer period of time. Apparently there is some form of immunologic memory associated with the immune response that enables the cells to recognize an antigen to which they have been previously exposed and immediately to begin producing an antibody as soon as the antigen is reintroduced.

All cells are able to form antibodies but almost all are formed in lymphocytes and plasma cells of the spleen, lymph nodes, bone marrow, and lymphoid tissue of the gastrointestinal tract. Cells that respond to antigenic stimuli by producing antibodies or by reacting directly with antigens are referred to as *immunologically competent cells*. The method in which the antigen is introduced into the body will determine which tissue will be most active in antibody formation; for example, if the antigen is breathed, the main production will begin in the lungs; if it enters through the intestinal mucosa, the major activity will be in lymph tissues. Once the cells have produced a specific antibody, they retain the capacity to continue producing the antibody for weeks, months, or years.

The relationship between the thymus gland and the immune response is of interest. Although composed mainly of lymphoid tissue, the thymus gland itself does not produce antibodies; rather, it forms lymphoidlike cells (lymphoblasts) that travel to the lymph nodes, spleen, and bone marrow, where they become precursors of plasma cells. In addition, the thymus produces a humoral factor that appears to provide the appropriate stimulus for the development of immunocompetent cells in these tissues. The thymus is apparently essential in relation to development of antibodies. In children born without a thymus, or when the thymus is removed from a fetus or infant immediately after birth, the ability to form adequate amounts of antibody in any of the antibody-forming tissues is significantly impaired.

The defense mechanism leading to immunity develops during the embryonic period; however, there is very little immunologic activity throughout fetal life. Although antigens are formed at various times in early development, the individual does not develop the ability to react to antigen until near the time of birth. Thus, by the time the individual's own antigens are formed, his immune mechanism does not recognize them as alien and therefore does not produce antibodies against his own proteins. This is sometimes referred to as *immunologic homeostasis,* or the capacity to distinguish between self and nonself. This tolerance normally lasts

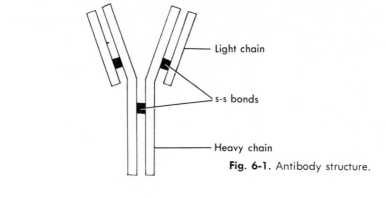

Fig. 6-1. Antibody structure.

Light chain

s-s bonds

Heavy chain

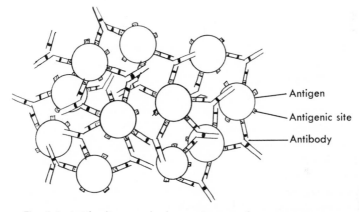

Antigen

Antigenic site

Antibody

Fig. 6-2. Antibodies attached to antigens to form an aggregate.

throughout the lifetime of the individual. It is uncertain whether this is due to functional immaturity of the cells or to humoral elements that render the process less efficient. Failure of this immunologic homeostasis can result in the serious autoimmune diseases to be discussed later. Closely related to self-recognition is *immunologic tolerance.* If an antigen is injected into a fetus during embryonic life the individual will always be tolerant of this antigen when his immune mechanism becomes functionally mature; it is recognized as part of the self.

Structure of immunoglobulins

Antibody structure is basically Y shaped. The immunoglobulin molecule is composed of two pairs of polypeptide chains—two light (L) chains and two heavy (H) chains—held together by three disulfide (-ss-) bonds. This organization allows the antibody to interlock with two antigens (Fig. 6-1). The base of the Y appears to be the point of attachment to the plasma cell; the two arms contain the binding sites. All immunoglobulins have the same basic structure but differ in the sequence of amino acids in the heavy chains of the molecule, which distinguishes one class from another. Fig. 6-2 illustrates the manner in which immunoglobulins attach to antigen to form an aggregate.

THE BLOOD GROUPS

The blood group systems have been one of the most valuable of all heritable traits in man in terms of their contribution to our understanding of the fields of genetics and anthropology. Their simple mode of inheritance and the variation in their frequencies among populations have provided some of the most useful genetic markers in man, which are helpful in family studies and the study of the distribution of genes in populations. The significance of blood group systems in medicine is well known, particularly in relation to blood transfusion and obstetrics.

The blood groups are genetically determined antigens of red blood cells (erythrocytes), and at least 15 different blood group systems have been identified, each determined by a separate gene locus. Unlike most of the inherited traits, the blood groups exist as *multiple alleles,* or *polymorphs,* which means that two or more genes occur at the same locus. Since each individual can carry no more than two genes at each locus (on the autosomes*), the combinations of genes from the different blood groups can produce extensive variations. For example, a blood group with three alleles (such as the ABO group) can combine to produce six different genotypes.

In the blood groups that have been identified the nomenclature is somewhat inconsistent (Appendix B). Some are designated by large and small letters (ABO, MNSs), others are named for the person in whom the antibody was first recognized (Kidd, Duffy). In addition to these there are a number of private blood groups (antigens that are found only in rare single kindreds) and public blood groups (antigens that are found in almost all persons). Each blood group is at a separate locus and is inherited independently of the other blood groups; for example, the ABO type of individual has no relationship to the Rh type as far as inheritance is concerned. Therefore, the transmission patterns of the various blood groups are easily discernible from each other.

The ABO blood group system

In 1900 Karl Landsteiner demonstrated that all human blood fell into four classes, based on the presence or absence of naturally occurring antigens on the red blood cells, and that these different antigens can be identified by means of antibodies. He found that the antibodies in the serum (the noncellular portion of the blood obtained after clotting) of certain individuals would cause the erythrocytes in these persons to agglutinate (clump together), while in others the red blood cells remain free. From these reactions he separated the antigens into the blood groups A, B, and O, which form the phenotypes A, B, AB, and O. The incidence of ABO groups in the population varies according to race and geographic area. In the North American white population the group that occurs most frequently is O (46%), group A is nearly as common (42%), group B is more rare (9%), and group AB is the least common (3%).

There is a reciprocal relationship between antigens on the red blood cells and antibodies in the serum. That is, group A persons, who have antigen A on their

---

*Only the Xg blood group is known to be carried on the X chromosome.

Table 6-1. Relationships of antigens and antibodies in the ABO blood group system

Blood group (phenotype)	Genotype	Antigens on red cells	Antibodies in serum
O	OO	none	Anti-A, anti-B
A	AA, AO	A	Anti-B
B	BB, BO	B	Anti-A
AB	AB	A and B	none

Table 6-2. Agglutination reactions in the ABO blood group system

Recipient Blood group	Antibodies in serum	Donor's RBCs O	A	B	AB
O	Anti-A, anti-B	–	+	+	+
A	Anti-B	–	–	+	+
B	Anti-A	–	+	–	+
AB	none	–	–	–	–

+ = agglutination
- = no agglutination

red cells, have anti-B antibody in their serum; group B persons have antigen B and anti-A antibody. Group AB persons have both A and B antigens and no antibodies against their antigens. Individuals with group O blood are free of both antigens A and B, but their serum contains both anti-A and anti-B antibodies. Since the genes for type O are recessive to both A and B, the genotype of a group A individual can be either AA or AO and type B can be BB or BO (Table 6-1).

ABO blood groups in transfusion. The discovery of ABO blood groups and the manner in which they interact have provided the information necessary to make blood transfusion reasonably safe for therapeutic purposes. The blood of any individual belongs to one of the four different types. Since antibodies in the serum of one group (blood type AB excepted) will produce an agglutination reaction when mixed with erythrocyte antigens of a different group, transfusion of whole blood is safe only between persons of the same blood group. This characteristic explains the severe reactions and deaths in persons that followed early attempts at blood transfusion. In a hypothetical situation, suppose that blood from a group B individual is introduced into a group A individual. Because the antibodies of the donor are diluted in the circulation of the recipient, the donor's serum (anti-A) will cause no ill effects in the recipient. However, the natural antibodies present in the recipient's blood (anti-B) will produce an agglutination reaction in the donor's red blood cells. The blood of the donor and the recipient is incompatible. Table 6-2 illustrates the expected agglutination reactions with each of the ABO blood types. To avoid the potential hazards related to transfusing whole blood, sensitive cross-matching is carried out to identify not only the major

ABO groups but any blood group factor that might cause an antigen-antibody response in the recipient.

Transfusion reaction is a serious and avoidable disorder based on the heritable characteristics of the blood group antigens. The blood groups A and B or Rh (D) (see the Rh blood group system) are involved in the greatest majority of reactions. Anti-A and anti-B are both naturally occurring antibodies, and antigens A, B, and Rh (D) are strong antigens to which antibodies are produced readily. When an antigen-antibody reaction takes place between incompatible blood groups the agglutinated donor cells become entrapped in the recipient's peripheral blood vessels. Usually within a period of hours (or sometimes days) the trapped cells degenerate and hemolyze (dissolve), liberating hemoglobin and other toxic substances into the circulatory system. The signs and symptoms displayed by the affected individual are caused by the blocking of blood vessels by agglutinated cells or result from the toxic effects of the increased concentration of bilirubin (from breakdown of the hemoglobin) and other substances. The most serious effect of transfusion reaction is acute kidney shutdown, which can begin almost immediately (with minutes) and continue until death from kidney failure.

Secretion of ABH antigens. Antigens A, B, and H* are found not only in red blood cells but may also be present in other tissues. Some persons (approximately 78% of the population) secrete them in body fluids in which they can be detected, such as, saliva and tears. The ability to secrete ABH antigens is determined by simple mendelian inheritance and depends on a single dominant gene *(Se)* independent of the *ABO* genes. Secretors have the genotype SeSe or Sese; nonsecretors are sese. The ability to secrete is inherited through the secretor gene; the type of antign secreted is determined by the *ABO* genes. Therefore, in order for a person to secrete type A antigen he must possess both genotypes and be either type AA, AB, or AO *and* SS or Se; to secrete type B antigens he must be type BB, AB, or BO *and* SS or Se.

ABO groups and disease. There is evidence to indicate that there is an association between some of the ABO blood groups and susceptibility to certain categories of disease. Most of the associations appear to involve diseases of the upper gastrointestinal tract where secretions of ABO blood group substances are present in large amounts. Cancer of the stomach, pernicious anemia (a form of anemia related to a decreased or absent production of mucoprotein by the stomach), salivary gland tumors, and diabetes mellitus all have an increased incidence in persons with blood group A compared with group O.[1]

Persons belonging to group O have a 35% to 50% greater likelihood of developing duodenal ulcers than persons in groups A, B, or AB; similarly, gastric ulcers are more frequent in group O individuals although to a lesser degree. An even closer association is found between nonsecretors and duodenal ulcer; nonsecretors are about 50% more likely to develop duodenal ulcers than secretors.

---

*The H antigen is the precursor substance from which antigens A and B are formed through the action of *A* and *B* genes. It is detectable in type O blood, which contains unaltered H substance.

Taking into consideration both of these factors, group O nonsecretors are about twice as apt to develop duodenal ulcers as group A secretors.

On the other hand, persons in group O seem to enjoy better health, based on a study of blood groups of active athletes over 40 years of age, volunteer soldiers selected for good health, and healthy persons over age 75. Compared with the general population, in these three groups of people there were significantly higher frequencies of group O blood.[1]

There has also been some association observed between ABO groups and some infectious diseases. A plausible explanation is that some infectious microorganisms possess antigens similar to the blood group antigens in human beings. It is theorized that, for example, a virus with an A-like substance might be partially neutralized by the anti-A antibodies in a group O person. Therefore, it would be expected that there would be a higher incidence of group A persons with such a virus infection. This has been found to be true with smallpox infection in an unvaccinated Asian Indian population living in primitive conditions. It has been postulated that perhaps this association between blood groups and disease might have played an important role in distributing the ratio of some blood groups during periods of severe epidemics.

The MNSs blood group system

The MNSs blood group system depends upon two sets of antigens: the MN and the Ss. The M and N are dependent upon a pair of codominant alleles that produce the phenotypes M, MN, and N from the genotypes MM, MN, and NN, respectively. However, the system is complicated by the Ss subdivision and other factors that will not be considered here. It appears that combinations of MN and Ss are inherited as units (for example, MS, Ms, NS, and Ns) to produce a variety of genotypes (MS, MNs, MNSs, Ns, and so forth). The precise nature of the inheritance is not clear but is probably the result of closely linked loci that remain together during crossover (p. 15). The major usefulness of the MNSs blood group system is in medicolegal problems of identification.

The Rh blood group system

The Rh blood group, named for the Rhesus monkey used in the experiments through which it was discovered, is one of major clinical importance. The Rh system consists of no less than eight gene complexes, and there is still controversy regarding the mechanism of inheritance. One faction (Fisher-Race) postulates a series of very closely linked genes that they call *C, D, E, c, d, e,* combined in two or more allelic forms. The other (Weiner and associates) assumes a single-gene locus with numerous allelic forms. These they give the base letters R and r with distinguishing superscripts: R^1, R^0, R', r^2, and so on. In this discussion the designation D and d of the Fisher-Race nomenclature will be used for simplicity. The D antigen is the strongest and most common Rh antigen and creates the greatest number of problems. Very simply, the *D* and *d* alleles produce three genotypes: DD, Dd, and dd. Phenotypically, DD and Dd are Rh-positive (contain D antigens); dd is Rh-negative (contains no antigens). The Rh (D) antigen is present in 85%

of the Caucasian populations, thus 15% are Rh-negative. Only 5% to 7% of Negroes are Rh-negative; in American Indians and the Mongoloid races the incidence is less than 1%.

Rh incompatibility. In the ABO blood group system the antibodies occur naturally. In the Rh system the person must first be exposed to the Rh (D) antigen, usually through transfusion of blood containing the antigen, before he will develop enough antibody to cause significant response. Ordinarily this creates no difficulty, but the second or third contact with the D antigen may produce a severe or even fatal reaction. For this reason blood typing includes the D factor as well as the ABO groups. A most serious immune reaction occurs during some pregnancies as a result of maternal-fetal incompatibility. Although the maternal and fetal circulation are distinctly separated, sometimes fetal red blood cells, with antigens foreign to the mother, gain access to the maternal circulation, which produces an immune response. The mechanisms and consequences of this interaction are discussed in the next section.

Blood group incompatibilities

Although most interactions between mother and fetus are not always well understood, the genetically determined immunologic incompatibilities that can lead to death or disease in the offspring are well known. *Erythroblastosis fetalis,* or *hemolytic disease of the newborn,* is an example of such a situation. The disease is characterized by progressive destruction of the infant's erythrocytes (hemolysis) and his efforts to form new ones (erythroblastosis). Until the discovery of the Rh factor the perinatal mortality from the disease was high, and many infants that survived were left with neurologic damage including mental retardation, hearing loss, and cerebral palsy.

INHERITANCE. Incompatibility between the mother and fetus can arise if the fetus inherits genes that produce proteins with the capacity to act as antigens in the maternal circulation. Hemolytic disease due to Rh incompatibility is the result of a mating between an Rh-negative (d) woman and an Rh-positive (D) man. The mother with a genotype dd can only transmit d genes to her offspring. The father with the phenotype D can be either DD or Dd. Through application of the principles of inheritance (Chapter 2) it can be predicted that if the father is homozygous (DD) any offspring from this mating will have an Rh-positive phenotype with the genotype Dd. If the father is heterozygous (Dd) one half the offspring from the mating will be heterozygous Rh-positive like the father and one half will be homozygous Rh-negative like the mother. It is impossible to tell from the phenotype whether Rh-positive individuals are homozygous or heterozygous.

DEVELOPMENT OF HEMOLYTIC DISEASE. Prerequisite to the development of hemolytic disease of the newborn is an Rh-negative mother (with no antibodies against Rh antigen) pregnant with a fetus whose blood cells contain the Rh-positive antigens inherited from an Rh-positive father. There is usually no difficulty with such a pregnancy until some of the fetal erythrocytes pass into the maternal circulation through breaks in the placental villi; then the mother's natural defense mechanism against foreign antigens responds by producing anti-Rh antibodies.

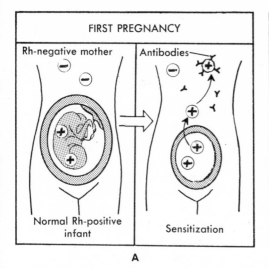
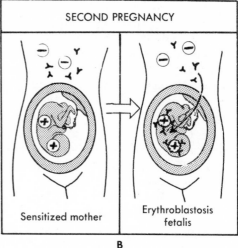

Fig. 6-3. Development of maternal sensitization to Rh antigens. **A,** Fetal Rh-positive erythrocytes enter the maternal system. Maternal anti-Rh antibodies are formed. **B,** Anti-Rh antibodies cross placental barrier and attack fetal erythrocytes.

These substances cross the placental barrier into the fetal blood stream, where they react with the Rh antigen on the fetal erythrocytes and destroy them, producing severe anemia (Fig. 6-3). Small numbers of fetal cells may cross the placental barrier during pregnancy, but the greatest stimulus occurs at the time of birth, particularly if the delivery is a traumatic one. Pregnancies with Rh-negative fetuses rarely create problems, as might be expected if the father is a heterozygote and transmits the recessive gene to the offspring or is himself Rh-negative. The first Rh-positive infant is usually unaffected; the hazard to the infant is greater in subsequent pregnancies. First pregnancies might be at risk if the mother has been previously sensitized by transfusion with blood containing Rh antigens or if she is highly susceptible to sensitization. There is wide variability in the ability of Rh-negative persons to produce anti-Rh antibody. To develop Rh antibodies in the Rh-negative mother, the Rh antigen must enter the maternal circulation and the mother must be capable of producing the anti-Rh antibodies.

CLINICAL MANIFESTATIONS. The disorder, which occurs in 80% to 85% of sensitized mothers, varies in severity from slight and transient symptoms to severe disease with lifelong disability or death. Often the severely affected infant is stillborn (hydrops fetalis). The symptoms that appear shortly after birth, usually within 24 hours, include anemia of varying degrees due to destruction of red blood cells, jaundice (yellow coloring of skin and sclera) from the accumulation of bilirubin as a result of excessive amounts of hemoglobin released from large numbers of hemolyzed erythrocytes, and hepatosplenomegaly (increased size of liver and spleen) created by an increased activity of these organs in an effort to compensate for the destructive process by excessive production of erythroblasts (precursors of red blood cells). Deposition of bilirubin in the brain cells can cause severe damage to the central nervous system *(kernicterus)* and result in mental retardation or cerebral palsy in the untreated infant.

TREATMENT. Symptomatic treatment of hemolytic disease in the newborn is aimed at decreasing bilirubin levels and replacing damaged erythrocytes with cells that cannot be destroyed by the antibodies. The usual therapy in the seriously affected infant is *exchange transfusion*. A calculated volume of the infant's blood is removed in small increments and replaced with fresh whole blood—Rh-negative, type O blood that has been typed and cross-matched with maternal blood. This therapy serves to reduce the amounts of circulating antibody, thereby decreasing hemolysis, and to increase the number of functional red blood cells. A significant decrease in hemoglobin or a serum bilirubin of 18 to 20 mg/100 ml of blood is indication for an exchange transfusion. An adjunct to transfusion therapy in some infants is the use of phototherapy (exposure of the nude infant to bright light). Although this effectively decreases the serum bilirubin levels, it has no effect on hemolysis and its long-term effects have not been evaluated.

Diagnosis of impending hemolytic disease is facilitated by the *Coombs* tests. The *direct* Coombs test detects antibodies already attached to the circulating erythrocytes of affected infants before the development of disease. This test is routinely carried out on babies of all Rh-negative mothers, with blood taken from the umbilical cord at the time of delivery. The *indirect* Coombs test detects the presence of anti-Rh antibodies in the maternal serum. Frequent monitoring of serum antibody levels (titers) during pregnancy in Rh-sensitized mothers forewarns of possible complications in the infant. If the maternal titers are sufficiently elevated to indicate the possibility of a distressed infant, diagnosis of erythroblastosis is confirmed by analysis of amniotic fluid bilirubin levels (see amniocentesis, p. 193). In carefully selected cases, when the gestational age contraindicates delivery and a fetus will otherwise die, intrauterine transfusion of compatible red blood cells is performed. The hazards associated with both therapies limit their widespread use.

The relatively recent development of anti-Rh immunoglobulin G (Rh$_0$GAM*) for the prevention of sensitization to the Rh factor has changed the outlook regarding hemolytic disease. When given to the mother before fetal antigens can stimulate the production of anti-Rh antibodies, Rh$_0$GAM provides temporary passive immunity for the mother against the immune response. Since sensitization almost always occurs after birth of the baby, immunoglobulin is given within 72 hours following delivery.† If this procedure is systematically carried out with all mothers who are at risk to develop Rh sensitization, the incidence of hemolytic

---

*Ortho Diagnostics, Raritan, N. J.

†There is reason to believe that the hormones of pregnancy have some immunosuppressive action during pregnancy that usually inhibits the immune response in the mother. Another explanation is related to the size and maturity of the immunoglobulins. Rh antibodies are of two types, IgM and IgG. The IgM variety is larger in size (usually the type that is produced following initial response to Rh antigens) and does not cross the placental barrier. The more mature variety, IgG (the type produced following repeated Rh-positive stimulation), is smaller and can readily traverse the placenta.

disease can be radically reduced or eliminated. The procedure cannot be applied to those women who are already sensitized to the Rh-positive antigen.

ABO incompatibilities

Hemolytic disease can also result when the major blood group antigens of the fetus are different from those of the mother. In such cases the maternal organism produces antibodies that can hemolyze the red blood cells of the fetus. Fortunately this happens in only a small percentage of ABO incompatible pregnancies. Although rare, hemolytic disease usually occurs when the fetus is blood group A or group B and the mother is group O. In this situation the naturally occurring anti-A or anti-B antibodies already present in the mother cross the placenta and attack the fetal red blood cells to produce hemolytic disease in the fetus. This type of blood group incompatibility is usually mild, and the infant is rarely born with severe anemia. The infant usually displays jaundice in the first 24 hours and may have hepatosplenomegaly. The laboratory tests and treatment for ABO incompatibilities are the same as they are for Rh incompatibility. Unfortunately, the more major antigens the mother lacks (that is, the more natural antibodies she possesses) the greater the risk of anemia in the infant and the higher the fetal loss.

More important, there seems to be an increase in the spontaneous abortion rate from ABO interaction between the mother and fetus during the early weeks of pregnancy. It seems that the naturally present antigens in a group O mother have a lethal effect on any pregnancy with a group A or group B conceptus. For instance, matings between a group O father and a group A mother will produce offspring in the expected ratios (all group A if the mother is homozygous, and 50% type A and 50% type O if she is heterozygous); in matings between a group A father and a group O mother there are fewer group A offspring than would be expected (all group A if he is homozygous and half group A and half group O if he is heterozygous) and the abortion rate is higher, also indicating early selection against group A concepti.

There is a protective effect against Rh hemolytic disease in ABO incompatible matings, however. Rh-negative women who are mated with men of incompatible ABO and Rh blood groups are much less likely to have Rh-affected children than matings with group O men; for example, a group A Rh-negative woman mated with a group B Rh-positive man. In this example, a possible explanation is that the red blood cells of a fetus with B antigens are destroyed by the maternal anti-B before they can induce the formation of anti-Rh antibodies by the mother; if the father is heterozygous for both the B and the Rh factors, one would find a higher than expected incidence of group O Rh-negative children.

Medicolegal applications of blood groups

The blood group systems have proved of value in determining relationships between a given individual and his parent: a parent and child may have been separated for years by some fateful event such as war or upheaval; an individual may claim to be a long-lost child of wealthy or prominent persons; rarely, newborn infants in a hospital nursery have been assigned to the wrong mothers. Some situa-

Table 6-3. Rules of inheritance of the ABO and MNSs blood group systems

Blood group of parents		Types of offspring	
Parent 1	Parent 2	Can produce	Cannot produce
ABO system			
O	O	O	A, B, AB
A	O	A, O	B, AB
A	A	A, O	B, AB
A	B	A, B, AB, O	none
B	O	B, O	A, AB
B	B	B, O	A, AB
AB	O	A, B	AB, O
AB	A	A, B, AB	O
AB	B	A, B, AB	O
AB	AB	A, B, AB	O
MNSs system			
M	M	M	N, MN
M	N	MN	M, N
M	MN	M, MN	N
N	N	N	M, MN
N	MN	MN, N	M
MN	MN	M, MN, N	none

tions involve crimes. The most common situation of doubtful relationship concerns disputed paternity. If all the persons involved are available for study the geneticist can often provide valuable assistance in such cases. Traits, such as the blood group systems, that consist of multiple alleles and depend on single-gene inheritance are well suited for analysis—a parent must possess a gene in order to transmit the gene to an offspring.

Unfortunately, in cases of disputed paternity it is not possible to say with certainty that a specific man is the father of a given child; however, it is frequently possible to say that a specific man could *not* be the father of a given child, depending upon the number of blood groups used. Such evidence is usually accepted in court. A man can be excluded from parentage if the child possesses an antigen that is lacking in both the mother and the putative father (for example, an AB child with an A mother and an O man) or if the child does not possess an antigen that he must have received from his assumed father (for example, an O child with an A mother and an AB man). On the other hand, several possible fathers might be implicated (for example, the father of a group A child with a group A mother could be type A, B, AB, or O). Table 6-3 lists the outcomes of matings of the ABO and MNSs blood group systems.

HISTOCOMPATIBILITY

Human organ transplantation has received international publicity in recent years. The technical problems involving most tissue and organ transplants are not insurmountable. Unfortunately, understanding of the problems related to rejection of the implanted tissue is not so far advanced. The primary reason for transplant failure is histoincompatibility (Gr. *histos,* tissue) between the donor and the recipient. Although it is a lifesaving characteristic, the body is a very hostile host to any alien material. A major goal in transplantation is not so much the surgical procedure but the lack of antagonism or toxicity between the different tissues that allows a graft to survive. The role of immunogenetics in transplantation is to elucidate the relationship of the transplant to the host and its genetic basis.

Histocompatibility antigens

There are antigens on cells other than erythrocytes. They are the *histocompatibility antigens* that are found on the surface of white blood cells, platelets, and most of the fixed tissues of the body (heart, kidney, liver, and so forth), and all carry the same antigens. The ABO antigens also occur on these tissues; therefore, they too can be considered as histocompatibility antigens. Rh and MN antigens, on the other hand, do not occur on cells other than erythrocytes. Histocompatibility antigens, like blood group antigens, are genetically determined and are also present in groups. Most of the investigations with these antigens have been done on laboratory animals, particularly the mouse, about which a fair amount is known. In man, it is estimated that there are probably 20 to 22 groups of antigens, and within every group there are possibly 15 to 30 antigens. The magnitude of the possible combinations gives some idea of the extent to which one individual differs from another.

The major antigen group, and the best known, has been designated *human leukocyte-A* (HL-A) and is the first human leukocyte locus to be identified. (There are other antigen systems known to be present, but they have not been identified as yet.) Each HL-A locus consists of at least two subloci (denoted HL-A$_1$, HL-A$_2$, HL-A$_3$, and so on) so closely linked that they usually remain together and segregate as a unit during meiosis. This unit, referred to as a *haplotype,* is designated by both subscripts, for example, HL-A$_{1,4}$ or HL-A$_{7,12}$. The genes for the HL-A antigens are codominant, and transmission follows the mendelian principles of inheritance: each offspring inherits one haplotype-bearing chromosome from each parent to determine his phenotype (Fig. 6-4). The HL-A system produces a wide variation of phenotypes in a population, which may prove a valuable tool for use in other areas such as medicolegal application and anthropology.

In order to minimize the likelihood of rejection, tissue transplantation requires the matching of as many major histocompatibility antigens as possible between donor and recipient. The process includes matching both the ABO blood group system on the erythrocytes and the HL-A system, using peripheral lymphocytes. HL-A antigens cause agglutination of white cells by the same mechanism that produces agglutination of red blood cells, which may be a factor in some transfusion reactions. Even if the donor and recipient are identical for these antigens there may still be problems. Another method used to determine donor suitability is a mixed leukocyte culture. It has been found that when differing HL-A antigens

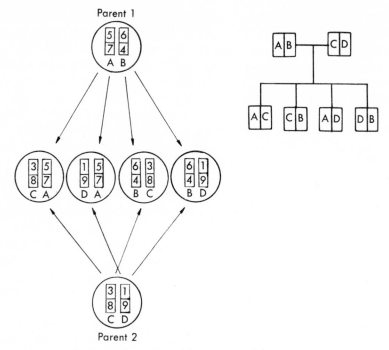

Fig. 6-4. Inheritance of histocompatibility antigens.

are mixed in culture they will mutually stimulate the cells to enlarge and divide. By inhibiting the activity of one group of cells with x-ray, the rate of DNA synthesis and mitosis can be measured in the other. The degree of genetic dissimilarity is reflected in the amount of cellular activity. Proliferation of lymphoid cells almost invariably accompanies an incompatible tissue graft.

Transplantation of tissues in man

Transplantation of tissues involves a genetically controlled and complex collection of cells. At the basis of the entire process is the concept of the uniqueness of self and the recognition of that which is nonself. The subtle mechanism that recognizes nonself and evokes a response has already been defined as the immune response. The primary goal in transplantation is the long-term survival of the grafted tissue. The means by which this is attempted are (1) securing tissues that are antigenically the same as the host's, (2) bypassing the host's immune mechanism, and (3) suppressing the host's immune mechanism.

Selection of donor tissue. Obviously, the closer that one individual is related genetically to another the more genes both will have in common and, therefore, the less likely will be the chance of producing an immune response. If the required tissue is such that it can be transferred from one site to another on the same person, such as skin or bone, no rejection will occur. This is referred to as an *autograft.* If the tissue needed is an entire organ, such as a kidney, there is only one donor with tissues genetically identical to those of the recipient—a monozygotic twin. In this type of transplant, an *isograft,* there should be 100% compatibility and total graft acceptance. Since monozygotic twins are relatively uncommon, the most frequent source of transplant tissue is from genetically dissimilar members of the same species. These are termed *allografts.* The nearer the relationship between the donor and the recipient the better the chance for either a delayed response or, in some kidney transplants, relatively long-term survival. Within families the same genotypes may recur; between families this is less likely. The best possible donor of an allograft is usually a sibling, who on the average has 25% of his genes in common with the recipient. The next best donor is a parent, then an uncle or aunt. Finally, a transplant between members of two different species, such as transplantation of a chimpanzee kidney into a human, is referred to as a *xenograft.* They are universally rejected.

The allograft reaction. The rejection of an allogenic graft by the host follows a predictable sequence of events. For the first few days the grafted tissue (for example, a skin graft) appears to take. The blood supply is continuous with that of the host and the skin looks pink and healthy. After the sixth to ninth day, depending upon the degree of incompatibility between host and donor, the grafted tissue begins to look inflamed from infiltration by lymphoid cells as they invade the graft bed. The blood supply shuts down, causing the graft to become swollen and cyanotic. The graft then dries and is sloughed off. This is the *primary* response.

Once the graft recipient has developed antibody to an implanted tissue and rejected it, a second graft from the same donor will be rejected more rapidly. The host has become sensitized to the donor tissue. This second rejection is called the *second set response.* These rejection responses are manifestations of the primary and secondary immune responses described earlier.

Bypassing the immune mechanism. Some tissues, primarily connective tissue, can be transplanted and either do not evoke the immune response or elicit a less marked reaction in the host. Corneal transplants will survive, possibly because the immune mechanism is not altered due to its peculiar lymph drainage system or, since the cornea contains no blood vessels, the antibodies do not gain access to the transplanted tissue. Bone grafts are not rejected, primarily because they do not replace the tissue but serve as a framework upon which new growth will take place. The nature of the transplanted tissue itself also influences the survival time. If the cells of the graft are nonnucleated cells, incapable of replication, they do not survive. A blood transfusion is an example of a short-term transplant. Red blood cells are, by their nature, doomed to short-term survival; even in the donor the life span of erythrocytes is only 120 days.

Suppression of the immune response. After the best possible tissue match is obtained for a transplant the survival time can be significantly lengthened by suppressing the immune response in the host. This can be accomplished in several ways. The immune response can be depressed with massive doses of ionizing radiation, total or local, which injures the entire blood-forming system. Drug therapy is usually employed, alone or in conjunction with radiation. The antimetabolites, such as methotrexate, actinomycin, 6-mercaptopurine, and azathioprine (Imuran), block protein synthesis and inhibit the growth of the rapidly dividing cells of the blood-forming bone marrow and lymphoid tissues. The drug is regulated to the lowest possible dosage to enable the immune response to remain partially functional. Corticosteroids (usually prednisone or prednisolone) are nonspecific immunosuppressants that are effective in suppressing the inflammatory process. All of these methods are not without hazard. Suppressing the immune response leaves the individual susceptible to a variety of viral and bacterial infections.

The newest of the methods used to suppress the immune response is antilymphocyte serum (ALS). To prepare ALS, lymphocytes from the proposed recipient are injected into horses, where they produce antibodies (antiserum). The antiserum administered to the organ recipient has been effective in improving the survival time of patients and transplants. Also, there seems to be fewer complications from infections with ALS-treated individuals.

Future tissue grafts may consist of cultured cells taken from the patient's own tissues, grown in the laboratory, and transplanted back into the individual. Such transplants would be immunologically safe, and when sprayed on raw surfaces the autologous cells would stimulate healing in any wound.

The placenta as a natural allograft. This remarkable organ not only survives for a relatively long period of time but performs vital functions efficiently during that time. Somehow the placenta escapes destruction by the maternal histocompatibility antigens. The phenomenon has excited interest for some time, but the precise mechanism is still unsettled. Controversy exists regarding whether there is a protective substance secreted by the early invading cells of the placenta or maternal immunologic unresponsiveness—some sort of antigenic stimulation by the fetus that causes the mother to produce immunoglobulins that coat the fetus and protect it from destruction by the mother's natural immune response mechanism.

AUTOIMMUNE DISEASES

The body normally exists in a state of immunologic homeostasis; that is, it is able to distinguish between endogenous (self) and exogenous (nonself) molecules. Although it has an aggregate of antigens and the capacity to react to antigens by forming antibodies, the body normally will not produce antibodies against its own antigens. However, under certain conditions there is evidence that the body does form antibodies against its own components, which is contrary to the long-accepted assumption that the immune response is directed only toward exogenous substances. This situation is termed *autoimmunity,* and the physical consequences are the *autoimmune diseases.* Unlike the ordinary antigen-antibody reaction, which is usually self-limiting (when the antigen is destroyed, the reaction ceases), autoimmune diseases are self-perpetuating and thereby produce lifelong disability.

As a group, autoimmune diseases seem to be more frequent in females and tend to arise in middle or late adulthood. There are several diseases classified as autoimmune diseases (for example, glomerulonephritis, myasthenia gravis, thrombocytopenic purpura), and there is evidence to implicate others (such as ulcerative colitis, scleroderma, sympathetic ophthalmia). The autoimmune disorders that appear to have a genetic etiology are Hashimoto's thyroiditis, rheumatoid arthritis, and systemic lupus erythematosus. The tissues involved in autoimmune responses may be *organ-specific* (limited to one organ) or *non–organ-specific,* or *systemic* (widely disseminated lesions).

Antigen-antibody basis of autoimmune disease

The way in which the antigen-antibody reaction is modified to produce such a response is probably due to either an alteration in an antigen (or the appearance of a new one) or the alteration in an antibody that will react with specific tissues.

Alteration in an antigen. Any alteration in a tissue will also result in a change in the antigen that it produces. The body, assuming the antigen to be foreign, responds by producing antibody against this tissue, termed the *target tissue* because it is the target for this specific antibody. Such an alteration could be caused by a somatic mutation of the genetic material or a viral infection. There is the possibility that some body substances have been isolated during embryonic life before immune tolerance has been established, and hence are not recognized as self. These substances behave as antigens if they are released from an organ through disease or injury. A constituent of the thyroid gland (thyroglobulin) could be such a substance. Some feel that sympathetic ophthalmia might be the result of this type of autoimmune mechanism. Severe injury to one eye may be followed days or weeks later by blindness in the other eye. Also, the development of immunity to protein released from muscle cells might make the cells less responsive to acetylcholine, producing the weakness of myasthenia gravis.

There is the possibility that autoimmunity may develop if a foreign antigen is similar in structure to some of the body's own proteins. For example, a strain of streptococci releases a toxin similar to protein found in heart muscle and valves and in the synovial membranes. A streptococcal infection will cause the body to produce antibody that may then attack these structures, causing in these particular tissues the symptoms and destructive lesions of rheumatic fever.

Alteration in an antibody. Another mechanism that might produce an auto-immune response is somatic mutation of a gene (or possibly a virus infection) that controls the production of an antibody. The mutation leads to the development of a *forbidden clone* (Gr. *klonos,* young shoot or twig) of cells. This new group of immunocompetent cells produces *autoantibodies* that react with normal body tissues and destroy them. This is the most popular explanation for the auto-immune phenomenon.

Systemic lupus erythematosus

Systemic lupus erythematosus (SLE) is not uncommon and, although it can be present at any age, usually first appears in the third or fourth decade. It occurs three to four times more frequently in females than in males. The symptoms are often preceded by an acute infection, stresses (such as surgery, trauma, or pregnancy), or after taking some types of drugs, particularly the antimicrobials and anticonvulsants. In almost all persons with SLE the serum shows high levels of abnormal gamma globulins that behave as antibodies.

The disease is believed to be an abnormal reaction of the body against its own connective tissues. In addition to such nonspecific symptoms as fever, malaise, and anorexia, the major manifestations of SLE consist of multiple cutaneous and systemic lesions. Cutaneous lesions occur in 85% of persons with the disease, and may appear alone or in conjunction with disorders of other organ systems. The characteristic "butterfly" rash that covers the malar regions and the bridge of the nose is the most striking feature of the disease. The patchy, erythematous rash often appears on the face, neck, and extremities, and may be quite bulbous and vascularized. It is frequently sensitive to sunlight. Autoimmune hemolytic anemia may be a dominant feature, and platelets and the peripheral leukocyte count are usually depressed. Common visceral sites include the pleura, synovial membranes, pericardium, liver, spleen, and kidney; renal involvement is frequently the terminal event.

SLE follows a lifelong course of remissions and exacerbations. The affected person is cautioned to avoid direct exposure to sunlight and any stressful event that might trigger an exacerbation of symptoms. The systemic manifestations of the disease respond to administration of corticosteroids, and the immunosuppressives have been used with some success.

The heritability is variable. There is some familial incidence of SLE but evidence suggests that environmental factors are equally influential. Approximately one tenth of the female relatives of typical cases are affected, and there is an increased incidence in monozygotic twins. Increased serum gamma globulins (characteristic of SLE) are found in a high proportion of relatives of persons with SLE. There is also a strong association between the individual with SLE and other of the autoimmune-type diseases, such as rheumatoid arthritis, in relatives.

Rheumatoid arthritis

The target organs in rheumatoid arthritis are the synovium, or lining of the joints. The inflammatory process in the joints produces edema and thickening of the synovial membranes, which eventually progresses to destruction of cartilage and bone and finally to ankylosis (an immovable joint). Usually more than one joint is involved and often it is the same joint bilaterally. The course of these processes varies from individual to individual in both rate and extent of involvement, from a single or occasional flare-up to severe disability and deformity. Exacerbations and remissions are characteristic.

Rheumatoid arthritis is associated with an autoantibody, called *rheumatoid factor,* that unites with gamma globulin of normal plasma. In other words, these autoantibodies are anti-antibodies acting on the tissues of the individual in which they were developed. Rheumatoid factor can be demonstrated in the plasma of persons with rheumatoid arthritis and also in individuals with SLE and their relatives. This familial incidence gives some strength to the suggestion that rheumatoid arthritis has a genetic basis, as do all the autoimmune diseases. The disease can be present at any age from infancy to old age although it most frequently appears in the fourth decade. It is three times more common in females than in males.

Treatment of rheumatoid arthritis is symptomatic. The preferred drugs, the salicylates (aspirin, Empirin, and so forth) have few serious side effects and have proved highly effective in suppressing the inflammatory process. Corticosteroids injected locally into painful joints provide pain relief, and given systemically often produce dramatic relief of symptoms. Neither of these drugs has any effect on the progression of the disease. Application of heat in the form of hot soaks, paraffin baths, and shortwave diathermy affords local relief.

Orthopedic devices (braces, splints, and so forth) may help prevent deformities, and physical therapy is an important part of the therapeutic regimen. Exercise movements, carried out slowly and carefully, are directed toward attaining the most complete range of motion possible for each joint. Wheelchairs, self-help devices (such as long shoehorns or thick-handled spoons) and facilities adapted to the disability afford the affected person as much independence as possible.

Hashimoto's thyroiditis

Hashimoto's thyroiditis, sometimes called autoimmune thyroid disease, is an example of an organ-specific autoimmune disease and is associated with the presence of circulating antibodies only to thyroid tissue. Enlarged thyroid (goiter) is the universal symptom, and hypothyroidism may also occur. Circulating autoantibodies to thyroglobulin can be detected in the serum of affected persons. There is infiltration of the gland by lymphoid cells, and it is believed that the antigen-antibody reaction in the thyroid gland leads to destruction of the functional thyroid tissue. This is an example where an isolated, or sequestered, substance is released and interpreted as a foreign antigen.

It has been found that Hashimoto's thyroiditis and other thyroid disorders (such as nontoxic goiter, myxedema, and thyrotoxicosis) are sometimes found in members of the same family. Also, autoantibodies are common in the sera of relatives of persons with Hashimoto's thyroiditis. The precise nature of the heritability is uncertain, but it is proposed that this high familial incidence of autoantibodies suggests that there might be a genetically determined failure to suppress the formation of antibodies against thyroid gland antigens that normally escape from the gland or that in these families there is a defect in the gland that allows an abnormal amount of antigen to escape.

Diagnosis is made on the basis of biopsy of thyroid tissue and the presence of high titers of thyroglobulin antibody. The therapy is thyroid hormone replacement.

INHERITED IMMUNOLOGIC DISORDERS

Some defects in the immune mechanism can be directly attributed to a single gene. Deficits in any of the proteins necessary for the immune process will alter the individual's ability to withstand infectious organisms. The outstanding manifestation of all these disorders is the large number and variety of infections.

Congenital agammaglobulinemia (Bruton's disease)

Congenital agammaglobulinemia is a disorder due to an extreme deficiency in the immunoglobulins IgA, IgG, and IgM. This is the most common form of the antibody deficiency diseases and is a result of a genetically determined abnormality of the lymphoid tissue. Apparently the mutant gene prevents the maturation of plasma cells. Plasma cells are absent from lymphatic structures and bone marrow, and the production of these cells cannot be stimulated by immunization as in normal persons. As a result, affected persons are susceptible to severe recurrent bacterial infections (including pneumonia, sinusitis, meningitis, and abscesses) and they do not respond to injected antigens (for example, diphtheria). They do respond normally to viral infection (measles, chickenpox), which has helped increase the understanding of the role of immunoglobulins in disorders of viral etiology.

The disease is transmitted as an X-linked recessive; therefore, affected individuals are almost always male. It is usually detected between 3 to 6 months of age after the transplacental immunity conferred by the mother has been depleted. These children require vigorous antimicrobial therapy for the recurrent infections and regular injections of gamma globulin to maintain a protective level of antibody. With this therapy, affected persons may now survive beyond early childhood but are subject to chronic infections and their effects, such as physical underdevelopment, and they frequently develop arthritic complications similar to those found in the autoimmune disorders.

Swiss-type agammaglobulinemia

A more severe form of agammaglobulinemia is the Swiss type. Persons with this disorder do not produce antibody of any kind and are also deficient in lymphocytes in both blood and tissues. The lymphoid tissues are absent and the thymus is poorly developed. These persons are highly susceptible to bacterial, viral, and fungal infections and they do not reject skin grafts. The inheritance pattern appears to be autosomal-recessive, and the affected individuals do not live beyond 2 to 3 years of age.

Wiskott-Aldrich syndrome

The Wiskott-Aldrich syndrome is an X-linked recessive immunodeficiency in which there is an impaired ability to produce antibody, especially to polysaccharide antigens (a common constituent of bacteria and fungi). There appears to be some defect in the ability to recognize or process this type of antigen. There are increased serum levels of IgA, normal IgG, but low levels of IgM. A constant feature is thrombocytopenia (decrease in blood platelets). Persons with this disorder are subject to eczema, hemorrhages from deficient platelets, and a wide variety of infections beginning in the first year of life that usually lead to death in early childhood, although some benefit is gained from plasma infusions. Also, there has been a high incidence of lymphatic malignancies associated with this disorder.

REFERENCE

1. Vogel, F.: ABO blood groups and disease, Am. J. Hum. Genet. **22:**464, 1970.

GENERAL REFERENCES

Bellanti, J. A.: Immunology, Philadelphia, 1971, W. B. Saunders Co.

Chung, C. S., and Morton, N. E.: Selection at the ABO locus, Am. J. Hum. Genet. **13:**9, 1961.

Crispins, C. G., Jr.: Essentials of medical genetics, New York, 1971, Harper & Row, Publishers.

Douglas, S. D., and Fudenberg, H. H.: Genetically determined defects in host resistance to infection; cellular immunologic aspects, Med. Clin. North Am. **53:** 903, 1969.

Fudenberg, H. H., Pink, J. R. L., and Wang, A.: Basic immunogenetics, New York, 1972, Oxford University Press, Inc.

Guyton, A.: Textbook of medical physiology, ed. 4, Philadelphia, 1971, W. B. Saunders Co.

Knudsen, A. G.: Genetics and disease, New York, 1965, The Blakiston Division, McGraw-Hill Book Co.

Porter, I. H.: Heredity and disease, New York, 1968, The Blakiston Division, McGraw-Hill Book Co.

Thompson, J. S., and Thompson, M. W.: Genetics in medicine, Philadelphia, 1966, W. B. Saunders Co.

7

Inheritance of common
diseases and disorders

The development of an individual throughout his life span is dependent upon a complex interaction between genetic components and environmental factors. The genetic constitution of the individual, present at conception and relatively unaltered during a lifetime, controls to some extent the way he responds to his environment. However, this genetic potential is continually influenced by the constantly changing environment. There is probably no situation in which one is not influenced in some way by the other. Both genetic and environmental factors interact to produce the phenotype, including those phenotypic characteristics interpreted as disease.

Genetic factors are probably involved in all disease processes, but in general only the rare diseases can be attributed to simple gene etiologies. In the more common diseases the extent of the genetic component varies from disease to disease, and although there is an increased incidence in families no clear-cut mode of inheritance can be identified. All diseases seem to fall on a continuous spectrum (Fig. 7-1).[8] On one end are those diseases determined entirely by the individual's genetic constitution, such as Down's syndrome, phenylketonuria, and muscular dystrophy. (Some, like phenylketonuria and galactosemia, are not on the extreme end because environmental influences in the form of dietary modification can significantly alter their course and outcome.) At the other end of the spectrum are the infectious diseases, such as tuberculosis, due primarily to environmental factors; but again there is evidence to indicate a decided genetic element in the susceptibility to the disease. Between these two extremes fall the bulk of common diseases with varying degrees of genetic influence. For example, there is a genetic component in the development of peptic ulcers, as evidenced by the familiar incidence, but probably more important in the causation of the disease are the effects of stress and worry.

136

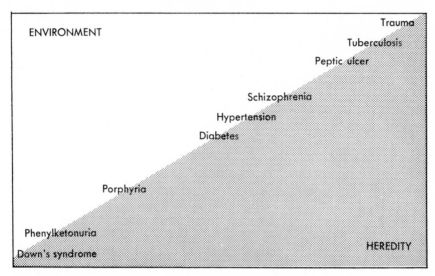

Fig. 7-1. Spectrum of diseases, indicating the relative importance of genetic and environmental factors.

To determine the significance of genetic factors in a given disorder several methods are utilized:[8]

1. *Family studies.* If the genetic component is important in the etiology, an increased incidence of the disorder in families should be demonstrated. However, environmental factors can play an equally significant role. For example, a dietary deficiency disease such as pellagra would be deceptive.

2. *Twin studies.* A higher concordance rate in genetically identical monozygotic twins than in dizygotic twins would tend to implicate genetic factors.

3. *Interracial comparisons.* Races show different frequencies for a number of genes. Variations in the incidence of a common disorder from race to race might indicate a strong genetic component. Again, environmental factors could be equally significant, since races are socially and geographically singular as well as biologically distinct.

4. *Component analysis.* Whenever there is a factor characteristic of a disorder, its presence or absence in families, races, twins, and animal homologues can provide valuable information; for example, glucose tolerance curves in relatives of diabetic patients and characteristic EEG patterns in persons with epilepsy.

5. *Blood group and disease association.* Some common disorders tend to occur more frequently than would be expected by chance in individuals with a specific blood group. Association with this simply inherited trait is evidence for a genetic factor—something in the physiologic makeup of the individual predisposes him to the disorder. For instance, there is an association between blood group O and the development of peptic ulcers.

6. *Animal homologues.* Many disorders in humans seem to have an identical counterpart in animals. Extensive studies can be carried out in animals to analyze the significance of genetic factors and the mechanisms of their contributions.

MULTIFACTORIAL AND POLYGENIC INHERITANCE

From the genetic point of view, diseases are frequently categorized according to the extent to which genetic factors are involved in their etiology. First there are those diseases due to a single mutant gene that are individually relatively rare but that collectively constitute a significant number of diseases, including cystic fibrosis, hemophilia, Huntington's chorea, and muscular dystrophy. Second are the chromosomal aberrations, which comprise another large group and include Down's, Turner's, and Klinefelter's syndromes. The disorders described up to this point have been primarily due to the large effects produced by a single gene—if the individual possesses the gene (in single or double dose) he is affected; if he does not have the gene he is unaffected.

The third classification encompasses those diseases that are *multifactorial* in their causation—those disorders in which a genetic susceptibility combined with the appropriate environmental agents interact to produce a phenotype that is interpreted as disease.[10] With multifactorial traits there is no clear-cut affected-unaffected classification, rather there is a *continuous variation* in the manifestations. The effects of each gene are small and additive. Most of the differences between normal human beings are due to continuous variation. The classic illustrations of multifactorial inheritance are stature, intelligence, and skin color. In addition, many common disorders including diabetes mellitus, schizophrenia, and hypertension are considered to be multifactorial.

Another term often used in relation to, and sometimes interchangeably with, multifactorial inheritance is *polygenic* (literally meaning many genes).* Polygenic refers to those genes that do not produce a large effect but are minor alleles at different loci whose combined effects produce a given phenotype, each making a small contribution to the total effect. When the trait under consideration is several steps away from the direct product of gene action, it is most likely that the trait produced is the result of the activity of many genes. This would include such traits as stature, closure of facial clefts, blood pressure, and finger ridge counts. In a sense, chromosome aberrations might be considered polygenic, since the loss or addition of chromosomes, or segments of chromosomes, that produces the phenotypic effect is composed of a large number of genes.[10] Both genetic and nongenetic factors are involved in multifactorial inheritance.

The distribution of polygenic traits is consistent with the bell-shaped curve (Fig. 7-2). In relation to height, the few tallest individuals are at one end and the few shortest at the other end, with the large majority of individuals of average height graded between these extremes. The nearer the center, or average, the larger the number of persons of that particular height. Statistically, only 5% of a given population represent the two extremes of the range and are significantly different from the mean. Stature, largely genetically determined, is felt to be the result of a large number of genes, each exerting a small effect (for example, some genes will act to increase height, some will act to decrease it) to produce a final result that

---

*There is not total agreement among authorities regarding the usage of polygenic and multifactorial inheritance. Some geneticists consider the term polygenic to mean the genetic component of multifactorial inheritance, while others use the terms synonymously.

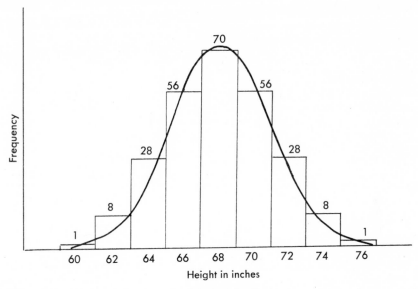

Fig. 7-2. Distribution curve for stature of 256 individuals. (From McKusick, V. A.: Human genetics, Englewood Cliffs, N. J., 1969, Prentice-Hall, Inc.)

is the sum of the single contributions. Sons of tall men will tend to be tall but usually not as tall as their fathers, and short men will not be as short as their short fathers. The sons' heights seem to fall halfway between the height of their exceptional fathers and the average height of the population as a whole. This is known as *filial regression* and is characteristic of polygenic inheritance. Regression assures that the overall population equilibrium is maintained. Although height is primarily attributed to the action of polygenes, other factors may distort the tendency: poor nutrition, assortative mating (tall persons who tend to marry each other), chronic disease, or abnormal conditions such as a dominant gene that produces a variation in height (dwarfism, arachnodactyly).

When the laws of inheritance are applied to polygenic characteristics, it is expected that relatives will have more genes in common. Therefore, there is an increased likelihood that these genes will be expressed more often when united with a similar combination of genes. If the gene(s) is very common, relatives will receive it from different sources; if the gene is a rare one, they will seldom inherit it. Table 7-1 shows the proportion of genes that relatives have in common. The more distant the relationship the fewer genes are shared. A very simple rule can be applied: for each step further away in relationship, the number of genes in common are divided by one half; that is, parent and child have half their genes in common (the child receives half of his genes from a single parent), a child and a grandparent have one fourth their genes in common, but monozygotic twins (with identical genes) have all their genes in common.

It has been postulated that there is a threshold effect in multifactorial inheritance: in the normal distribution curve there is a point beyond which the sum of

Table 7-1. Proportion of genes in common in various relationships

Relationship	Proportion of genes in common
First-degree relatives	
Parent, child, sib	½
Second-degree relatives	
Grandparent, grandchild, uncle, aunt, nephew, niece, half-sib	¼
Third-degree relatives	
First cousins	⅛
Second cousins	1/32

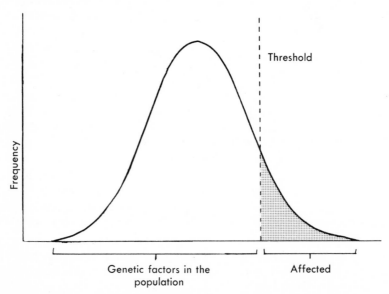

Fig. 7-3. Normal distribution curve illustrating the threshold concept of polygenic inheritance.

the accumulated genes produces the disease or disorder. In the general population this threshold is situated at the extreme of the curve so that affective individuals constitute a very small proportion of the total. In families in which a condition is known to occur there are a larger percentage of genes that combine to produce the disorder. The threshold in these families is found to be nearer the mean, and therefore the number of affected persons is proportionately larger than that in the population as a whole (Fig. 7-3). In such families there is an increased prevalence of a disorder to the extent that in first-degree relatives (brother, sister, son, daughter) the frequency may be 3 to 15 times that in the toal populaion. More than one affected family member indicates a greater number of multifactorial traits in common. However, socioeconomic differences or seasonal distribution might suggest environmental influences.

This chapter is concerned with diseases that cannot always be attributed to a specific genetic factor. These consist of the more frequently seen congenital defects, the more common diseases, the genetic influence on the development of cancer, and the relationship between genetic factors and certain drug reactions.

COMMON CONGENITAL DEFECTS

The influence of exogenous factors in the causation of congenital defects has been covered to some extent (Chapter 5). There are numerous specific defects that are seen with increased frequency in the population and that have a tendency to show an increased incidence in some families, indicating a heritable predisposition to these conditions. Most are consistent with the multifactorial/threshold concept, but there appears to be a stronger genetic component in clefts and positional foot defects than in CNS malformations and congenital heart disease.

Congenital dislocation of the hip[16]

Congenital dislocation of the hip (CDH) usually begins as a dysplasia (abnormality of development) rather than an actual dislocation. The dysplastic hip joint consists of a flat acetabular roof and an underdeveloped femur head. This dysplasia, which makes up 98% of cases of CDH, produces an insecure anchoring of the femur within the acetabulum (hip socket) and predisposes to dislocation after birth. The cause of this dysplasia is not known. During the first trimester of pregnancy there are numerous chances for maldevelopment. It may be due to delayed rotation of a limb bud, or muscle innervation during embryologic development, or malposition of the fetus in the uterus during the later months of pregnancy. In some instances it may be due to generalized and abnormal relaxation of the joint ligaments, possibly from hormonal factors. Hormonal changes in the mother toward the end of pregnancy that produce relaxation of the pelvic joints to facilitate birth may be a predisposing influence, especially in female infants. A frank breech position (buttocks first with legs extended) at birth and swaddling after birth certainly are factors in any susceptible infant. Any condition that adversely affects the development and relationship of the femoral head to the acetabulum may cause dysplasia and subsequent dislocation. Untreated CDH produces a shortened leg on the affected side and an abnormal waddling gait, accompanied by lordosis and outward rotation of the affected leg.

HEREDITY AND INCIDENCE. The incidence of CDH is about 1 in 500 to 1,000 births and occurs more frequently in females than in males (1:7). The disorder occurs 25 to 30 times as often in first-degree relatives as in the general population. Concordance in monozygotic twins is 40% and 3% in dizygotic twins, which suggests that genetic factors play a role in the etiology of CDH. It is also common to find hip dysplasia without dislocation in relatives of affected persons (without the female predisposition); therefore, the development of dislocation is apparently related to other factors. One fourth of cases involve both hips, and when only one is involved the left hip is affected 3 times more often than the right. CDH is frequently associated with other conditions, such as spina bifida. The risk of a second affected child is approximately 5%, or 1 in 20 in families with normal parents and one affected child.

Factors that are interesting but unexplained are that the incidence in first-born children is double that for second- or third-born, and the frequency among children born in winter is double that of children born in summer. There is some relationship to the development of dislocation and the methods of handling infants. Dislocation occurs less frequently in the Orient, where mothers traditionally carry their infants on their hips. Among the Lapps and some American Indian tribes, where infants are heavily swaddled or tied to cradle-boards, the incidence is particularly high.

DIAGNOSIS AND TREATMENT. X-ray examination is helpful, but diagnosis can usually be made by inspection. An early and obvious abnormality is a shortened leg on the affected side and unequal skin folds in hips and thighs. There is an extra fold on the affected side viewed anteriorly or posteriorly. There is restricted outward movement (abduction) of the affected hip, and manipulation reveals telescoping of the joint and a palpable or audible click as the head of the femur leaves the acetabulum.

Treatment consists of maintaining the hip in abduction (frog position) with several diapers, which is usually sufficient to produce desired results in mild cases. If this is not enough to encourage proper readjustment with growth or if the hip is severely dislocated it may require casting or even surgical intervention.

Clubfoot

Clubfoot is a general term used to describe a deformity in which the foot is twisted out of shape or position. The most frequent deformity (approximately 95%) is the *talipes equinovarus* variety in which the foot is pointed downward and inward in varying degrees of severity. Clubfoot may occur as an isolated defect or in association with other disorders such as arthrogryposis (a generalized immobility of the joints), congenital dislocation of the hip, cerebral palsy, or spina bifida. The feet are usually in the equinovarus position during early development but gradually assume the normal position in the final months of prenatal life. Persistence of this position may be due to a genetic predisposition, abnormal intrauterine position, or other adverse prenatal influence. It may also be a manifestation of a neurologic problem such as spina bifida.

The frequency of clubfoot in the general population is about 1 in every 700 to 1,000 live births, and boys are affected twice as often as girls. There is a 35% concordance in monozygotic twins as opposed to a 3% concordance in dizygotic twins, which indicates a genetic component. The frequency in first-degree relatives is 3% which is about 25 times the population incidence. The risk of having a second child with clubfoot is approximately 1 in 20 but varies with the sex of the sibling. If the affected person is male, the risk to a male sib is 1 in 42, but to a female sib is virtually zero. If the affected person is female the risk to a male sib is 1 in 16; to a female sib, 1 in 40.

TREATMENT. Treatment is begun early and can often be effected with shoes fitted to a crossbar that maintains the feet in an outward angle (Denis Browne splint) to create overextension, aided by the child's normal kicking. Wedge casts are sometimes applied. Only severe cases require forcible manipulation or surgery. Treatment is a prolonged process that extends from birth to puberty since there is a tendency for the deformity to recur.

Cleft lip and palate[6, 14]

Cleft lip with or without cleft palate, or CL(P), apparently has a different and distinct etiology from cleft palate (CP) alone.

Cleft lip with or without palate. CL(P), the most common of the facial anomalies, results from failure in fusion of the maxillary process with the nasal processes of the frontal prominence that normally occurs about the sixth week of gestation. Various degrees of incomplete fusion will produce the variations in degrees of cleft, from a notch in the vermillion border of the lip to complete separation extending to the floor of the nose. It may or may not involve the primary palate. Fusion of the secondary palate (roof of the mouth) takes place at a later stage. Clefts may be unilateral or bilateral, complete or incomplete. The precise cause of this developmental failure to fuse is unknown. Perhaps there is a developmental delay in the inward movement of the maxillary process or the nasal process develops too slowly and the gap never closes. The processes may meet but fail to fuse, or they may break apart. It might even result from a generalized developmental instability in the embryo.

CL(P) occurs in about 1 in 600 to 1,000 births. It is twice as frequent in Caucasians as in Negroes and more common in Japanese than in Caucasians. The concordance rate in monozygotic twins is 40% and 8% in dizygotic twins. Affected males outnumber females, particularly in the more severe defects; that is, greater for CL(P) than CP. Where the cleft is unilateral, about two thirds are on the left side; and an associated cleft palate is found more often with bilateral than with unilateral cleft lip.

There are no less than 50 recognized syndromes that include CL(P) as a feature, some due to mutant genes, others due to chromosomal aberrations (trisomy D and E), and vary rarely to an environmental agent. It has been known to occur in syndromes caused by teratogens such as rubella infection or thalidomide ingestion, and there may be an increased incidence with parental age. There is a high mortality during the newborn period in these cases, probably because of the multiple anomalies associated with most of these syndromes.

The incidence in relatives is consistent with the concept of multifactorial inheritance. The risk for a subsequent affected child is approximately 4% if the parents of a child with CL(P) are normal and have no affected relatives. After one child with CL(P), the risk for a second affected child increases to 9%—the parents can be assumed to carry more of the genes predisposing to CL(P) and are, therefore, nearer to the threshold. If one parent has CL(P) the risk for siblings increases from 4% after one affected child to 14% after two affected children. The risk does not seem to increase when parents are related or when there is an affected relative other than a parent or sib.

Cleft palate without cleft lip. In isolated cleft palate the defect appears to be related to the secondary palate, which is developmentally distinct from the primary palate. The secondary palate closes later in development, at about 8 to 9 weeks of intrauterine life. The palatine shelves in the embryo must migrate from a vertical position at the side of the tongue to a horizontal one above the tongue, there to fuse and form the roof of the mouth. If there is a delay in this movement the remainder of development proceeds, and the palates never become fused. The degree of severity varies.

The incidence of CL is lower than CL(P), with a frequency of about 1 in 2,500 in Caucasian populations but is lower in Negroes and higher in Japanese. It usually occurs as an isolated defect and is more common in females than males. There appears to be no relationship between CL and CL(P). Relatives of individuals with CL have an increased incidence of CL but not CL(P) and vice versa. The concordance in monozygotic twins is 24%; in dizygotic twins, 10%. Normal parents with an affected child have a 2% probability of producing a similarly affected child; and an affected parent has a 6% likelihood of having an affected child. When an affected parent has one affected child the risk for subsequent children with CP is about 17%.

There is one rare form of CP, Van der Woude's syndrome, that is associated with small pits on the lower lip (openings of extra salivary glands). It is inherited as autosomal-dominant and conforms to mendelian principles.

DIAGNOSIS AND TREATMENT. The defect is readily apparent. Corrective treatment for these disorders is surgical closure of the clefts. Prior to surgical repair, an important aspect of child care is a special feeding technique to compensate for the inability to suck in the normal fashion. The cleft lip is usually repaired as soon as the infant can tolerate surgery, but the palate is not closed until 2 to 3 years of age in order to take advantage of the normal growth processes. Secondary treatment by way of orthodontia and speech therapy is an important aspect of long-term care of these children.

Central nervous system malformations

Malformations of the central nervous system constitute the largest group of anomalies consistent with a multifactorial/threshold etiology. Two of these defects, anencephaly and spina bifida, occur in association with one another more often than would be expected by chance, suggesting a common origin. The various central nervous system defects (anencephaly, hydrocephalus, and spina bifida) may alternate in sibships, which also tends to bear out the theory of common origin.

Anencephaly. One of the most common of the central nervous system defects, anencephaly is a developmental anomaly characterized by absence of the cranial vault (upper skull) and a brain that merely consists of a vascular mass. It is apparently due to improper closure of the neural tube during early embryonic development, with secondary degeneration. It occurs more than twice as often in females as in males, probably due to a selective loss of boys with anencephaly by abortion in early development. Estimates of the incidence vary, but it appears to be approximately 1 in 1,000 births. There seems to be considerable variation in geographic location, socioeconomical level (more frequent among the poor), season of the year (increased incidence in the autumn and winter), and age of the mother (an increased incidence in first-born of young mothers and in mothers over 40 years of age). This disorder is fatal.

Spina bifida. The neural arches of the vertebrae normally unite about the twelfth week of gestation. When they fail to unite, leaving the spinal cord and its lining (meninges) unprotected and protruding through the defective vertebral arch, the result is a defect known as spina bifida. This failure to close produces a cyst or sac, usually in the lumbosacral region of the spinal column, containing cerebrospinal fluid, spinal cord, and nerve roots *(myelomeningocele)*. In the less severe manifestation, *meningocele,* the contents of the sac contain only meningeal tissue and spinal fluid, and there is little or no nerve involvement. Least serious of these spinal cord defects is *spina bifida occulta,* in which there is a defect in the spinal column but no protrusion of underlying soft tissues. The degree of deformity and the nerves involved determine the extent of neurologic deficit in any form of spina bifida. There are few disabilities related to the occult variety, but individuals with spinal cord and nerve root involvement are usually deficient in movement and sensation in the areas supplied by the affected nerves.[16]

Spina bifida as an isolated defect occurs less frequently than anencephaly, but the methods of recording spina bifida lead to some confusion regarding its actual incidence. Because it is so often associated with anencephaly or hydrocephalus it is sometimes grouped with these malformations, and the data are not clear-cut. The disorder usually occurs sporadically but there seems to be a familial tendency in many instances. Dislocation of the hip is a physical defect often associated with myelomeningocele.

TREATMENT. Surgical repair of the sac is accomplished as soon as possible. Until surgery can be attempted it is essential to prevent infection in the thinly covered meningocele that can rapidly spread throughout the central nervous system. Long-term care is needed to cope with the associated physical problems related to deficient nerve supply to bladder and bowel muscles and lack of sensation in the lower extremities. These children require bowel and bladder training, and are prone to frequent urinary tract infections with their concomitant hazards. Braces and other orthopedic appliances are helpful, but care must be taken to prevent pressure sores in the insensitive lower extremities.

Hydrocephalus. Hydrocephalus is characterized by an abnormal accumulation of cerebrospinal fluid (CSF) within the cranial vault. When the condition occurs in infancy it produces enlargement of the head and dilatation of the ventricles, accompanied by atrophy of the brain substance, mental deficiency, and convulsive seizures.

Various mechanisms have been described to explain the development of congenital hydrocephalus. Overproduction of CSF has been established as a cause in some instances and defective absorption has been suggested, but the most frequent cause is considered to be a block in the normal circulation of CSF. The entire brain and spinal cord is bathed in CSF. Normally this fluid is secreted, circulated, and absorbed to maintain a constant volume within the closed system. The classic concept assigns fluid secretion to the choroid plexuses situated in the hollow portions of the brain (ventricles) and absorption to the arachnoid spaces and villi surrounding the brain. The fluid travels between the four ventricles through narrow channels (foramen of Monro, aqueduct of Sylvius, and foramina of Magendie and Luschka) and to the spinal cord through an opening in the base of the skull, the foramen magnum. A block in the circulation at any of these points will produce an accumulation of CSF behind the obstruction, creating increased pressure against the brain. Malformations within the system are the most frequent cause of congenital hydrocephalus; the most common is the Arnold-Chiari deformity. This abnormality is produced by a downward displacement of portions of the cerebellum and medulla oblongata through the foramen magnum, which obstructs the flow of CSF from the ventricles to the spinal canal and the space surrounding the brain. The Arnold-Chiari malformation is frequently associated with myelomeningocele. Obstruction or obliteration of the other foramina, especially aqueduct stenosis, is often the causative defect.

INCIDENCE. Hydrocephalus occurs in approximately 1 to 1.5 in 1,000 births, but when associated with spina bifida the rates are between 1.5 to 3 in 1,000 births. In cases where spina bifida is absent there is an excess of males with hydrocephalus; where spina bifida is part of the anomaly there is an excess of females. In some families hydrocephalus is inherited as autosomal-recessive, and a form of aqueduct stenosis occurs as an X-linked recessive trait. Environmental factors known to cause hydrocephalus include maternal toxoplasmosis, cytomegalic inclusion disease, and syphilis, or various intrauterine infections of the fetus.* In these the agent is not likely to be operative in subsequent pregnancies.

DIAGNOSIS. Hydrocephalus is diagnosed at birth by observation in many instances, although often the defect is not recognized until later childhood because of the insidious progress of the increased pressure. Daily head circumference measurement in suspected cases may indicate the condition, but verification is made by x-ray examination using air or contrast media.

TREATMENT. Treatment is directed toward relieving and arresting the increased intracranial pressure, although numerous shunting procedures have met with varying degrees of success. Most utilize the basic principle of diverting the CSF from the ventricles to some area of absorption such as the peritoneum, thoracic duct, a ureter, or the atrium of the heart (the area preferred by most neurosurgeons).

Recurrence risks.[14] Except for hydrocephalus known to be due to a single gene (for example, X-linked hydrocephalus), a second affected offspring will be less likely to have the same central nervous system defect but can have *any* of them (anencephaly, spina bifida where the brain involvement is unknown, or spina bifida with Arnold-Chiari deformity). The risk to a subsequent sib is between 1 in 25 to 40, and greater if the mother has had several abortions. After two affected children the risk is 1 in 10 to 15; the risk after three central nervous system defects is very high at 1 in 3 to 5. For spina bifida not associated with the Arnold-Chiari malformation the risk is probably 1 in 40 to 50.

---

*Infection is the cause of obstructive hydrocephalus in about 50% of cases that develop after birth.

Pyloric stenosis

Hypertrophic pyloric stenosis is not actually a congenital malformation but a common functional anomaly present at or soon after birth. The circular musculature of the pyloric sphincter, through which gastric contents pass into the duodenum, becomes hypertrophied (enlarged) to the extent that the opening becomes occluded. It may be present at birth, but manifestations are rarely noticeably until 2 or 3 weeks of age when the infant begins to vomit during or shortly after feedings. The vomiting becomes progressively more marked and projectile in character, and if untreated there is a high mortality from pyloric stenosis. Surgical relief of the obstruction by the Ramstedt operation (a longitudinal splitting of the pyloric muscle) produces excellent results.

INCIDENCE. Pyloric stenosis occurs in about 3 in 1,000 live births, and about 80% of affected infants are males. It is seen less frequently in Negroes and Orientals than in Caucasians. The concordance in monozygotic twins is 30% to 50%, which suggests that the etiology must be due to the effects of both genetic and environmental influences. Some environmental factors that were identified in an extensive study[2] may be of importance: (1) pyloric stenosis is more common in first-born children, especially in infants where the onset of symptoms was in the third week or later; (2) the onset of symptoms was later in infants born in the hospital than in those born at home; and (3) the onset of symptoms was later in children fed every 4 hours than in those fed every 3 hours. These factors suggest that some aspect of feeding may postpone or even prevent the development of symptoms.

The predominant theory regarding the etiology of pyloric stenosis is that the predisposing genotype has two components: a common dominant gene and a sex-modified multifactorial background (that is, a higher threshold in females than in males). In studies of survivors and the incidence of the disorder in their families, it has been found that males who had pyloric stenosis in infancy had 7% affected sons and 1.5% affected daughters, while females who had had the disorder bore 24% affected sons and 4% affected daughters. Although females are four times less susceptible than males, those who were affected had four times as many affected children. Apparently some sex-dependent factors inhibit the manifestation of this anomaly in females. Since they must possess many of the responsible genes (a strong genetic liability) they are more apt to pass them to their offspring. There is a strong impression that the multifactorial component may be related to general body musculature.[2, 4]

Congenital heart malformations

Congenital malformations of the heart comprise a large and varied group of developmental defects, and there is wide variation in the severity of almost every defect. Some occur as features of syndromes due to chromosomal abnormalities (all autosomal aberrations and Turner's syndrome) or single-gene disorders (Marfan's syndrome, Holt-Oram syndrome). Others can be attributed to environmental agents. Rubella infection in the mother during early pregnancy has been proved to be a cause, as well as radiation and some drugs (thalidomide). Cardiac anomalies are frequently associated with neural tube defects and infants with any severe multiple malformations. Only a small proportion of congenital heart malformations can be definitely attributed to a genetic etiology. Its occasional appearance in several generations suggests such a cause, but sporadic cases might also be genetically determined. The high mortality in early infancy or before the reproductive age may represent a lethal or semilethal dominant gene, but without transmission no pattern can be identified. Perhaps with the increased survival rate made possible by newer surgical techniques the heritability of many cardiac anomalies can be analyzed.

Almost all of these large and varied malformations interfere in some way with normal flow of blood through the chambers of the heart and great vessels. In many abnormalities blood is short-circuited through defects in the heart or great vessels, which diverts the blood toward one of the other of the two main blood circuits: pulmonary or systemic. When blood is shunted from the right side to the left without passing through the pulmonary circulation to pick up oxygen, the individual displays varying degrees of cyanosis. On the other hand, blood shunted from the left side to the right is sufficiently oxygenated, but a larger than normal volume of blood is delivered to the lungs, causing congestion with all its associated complications, such as congestive heart failure or pneumonia. The turbulence created by blood flowing through these abnormal openings produces the murmurs characteristic of specific defects.

Most congenital heart malformations appear to be due to interference with or arrest of the normal developmental process, and represent persistence of a state that is normal at some stage of ontogeny. The heart evolves from a single chamber, which becomes divided into four by gradual invagination of septal walls, and a single vessel that twists and separates into the two major vessels: the aorta and pulmonary artery. The more common of the congenital heart defects are listed in Table 7-2 with their incidence and a description of the defect and associated disorders.

Risk estimates for cardiac anomalies due to a single gene or chromosomal aberrations are given in relation to the total syndrome. Isolated cardiac defects present a more difficult problem. As a whole, malformations of the heart do not show a strong genetic susceptibility, and the overall risk to sibs of an affected child is about 1 in 50. The probability of a sib being affected with any defect varies according to the specific malformation in the proband: 0.3% for ventricular septal defect to 2% for pulmonic stenosis. The frequency of a disorder in parents of affected individuals is low, but little is known of frequencies in offspring of affected parents. The concordance is usually low for both monozygotic and dizygotic twins, which tends to increase the importance of environmental factors in the etiology. A significant postnatal environmental factor in patent ductus arteriosus is hypoxia. The evidence of this anomaly is more common in populations living at high altitudes than those living at sea level.

COMMON DISEASES[3, 13]

Multifactorial inheritance similar to that for congenital malformations has been implicated in a number of common diseases. In many there is a higher incidence in families than would be expected by chance, but no specific mode of inheritance can be identified. In most, environmental factors also play an important role.

Table 7-2. Summary of congenital heart defects*

Anomaly	Male to female ratio	Definition	Percent-age of incidence of CHD in infants†	Disorders associated with increased incidence
Ventricular septal defect (VSD)	1:1	Abnormal opening between the ventricles	28.3	Down's syndrome
Patent ductus arteriosus (PDA)	1:3	Persistence of the fetal connection between the aorta and pulmonary artery	12.5	Rubella syndrome
Atrial septal defect (ASD)	1:3	Abnormal opening between the atria	9.7	Holt-Oram syndrome, Down's syndrome
Coarctation of the aorta	4:1	Stricture in the thoracic aorta	8.8	Turner's syndrome
Transposition of the great vessels (TGV)	3:1	Reversal in the origin of the great vessels—the aorta from the right ventricle and the pulmonary artery from the left ventricle	8.0	Diabetes or pre-diabetes in the mother
Tetralogy of Fallot	1:1	Ventricular septal defect and pulmonic stenosis with right ventricular hypertrophy and aorta overriding the ventricular septum	7.0	Thalidomide ingestion
Pulmonic stenosis	1:1	Narrowing of the opening into the pulmonary artery	6.0	
Aortic stenosis	4:1	Narrowing of the opening into the aorta	3.5	
Truncus arteriosus	1:1	One large single vessel leaving the base of the heart	2.7	Thalidomide ingestion
Other			13.5	

*Data from Perloff, J. K.: The clinical recognition of congenital heart disease, Philadelphia, 1970, W. B. Saunders Co.; and Warkany, J.: Congenital malformations, Chicago, 1971, Year Book Medical Publishers, Inc.
†Data from Campbell, M. In Watson, A., editor: Paediatric Cardiology, London, 1968, Lloyd-Luke, Ltd., Chapter 5.

Diabetes mellitus

The name diabetes, derived from the Greek term that means "to run through," describes the long-recognized syndrome of polyphagia (voracious eating), polydypsia (excessive thirst), and polyuria (large volume of urine). Mellitus means sweet, as sugar or honey. It is one of the most prevalent metabolic diseases in the world, and although there is no cure it is a condition for which there is a remedy.

Diabetes mellitus is a disorder involving primarily carbohydrate metabolism that shows a strong genetic component but is clearly influenced by environmental factors. The disease is characterized by a deficiency or diminished effectiveness of the hormone insulin. Without insulin, except for a very few specialized cells, the body is unable to metabolize carbohydrates adequately, which leads to increased blood glucose levels (hyperglycemia) and the appearance of glucose in the urine (glycosuria). Consequently, the major source of energy must be derived from metabolism of lipids. The products of this fatty acid metabolism (ketosis) are increased in the blood stream (ketoacidosis) and excreted in the urine (ketonuria). The excess acid results in renal loss of sodium and potassium, the increased glucose in the blood stream creates an osmotic diuresis (increase in urine) that results in increased thirst. The ever-widening effects of this defective metabolism extend to other areas. Body-building (anabolic) processes are halted or diminished, and there is an increase in the breakdown (catabolic) processes. Protein is mobilized in an effort to secure energy (gluconeogenesis), resulting in loss of body weight. Almost all areas of the body are affected by these metabolic adjustments, and the long-term effects of diabetes are a marked susceptibility to generalized vascular disease.

Heritable diabetes is classified into four major categories according to the extent of clinical manifestations: (1) *prediabetes,* the stage preceding any clinical signs; (2) latent chemical, or *stress diabetes,* an abnormal glucose tolerance test under stress (for example, illness or pregnancy) that at other times is normal; (3) *chemical, latent,* or *asymptomatic diabetes,* an abnormal glucose tolerance test but a normal fasting blood glucose in the absence of stress; (4) *overt,* or *clinical diabetes,* symptomatic diabetes, further divided into (a) *juvenile* (growth-onset or ketosis-prone) diabetes, which occurs in persons less than 20 years of age, develops rapidly, and is usually more severe and unstable; and (b) *adult* (maturity-onset or ketosis-resistant) diabetes, which develops more slowly and is usually milder and more stable.

BASIC DEFECT. The precise nature of the basic defect and the mechanism by which insulin acts are not clear. Both juvenile and adult diabetes respond to administration of insulin. In juvenile diabetes there is a reduction in islet cell mass or destruction of islets, resulting in a deficiency of insulin. In maturity-onset diabetes there are viable islet cells and some insulin in the pancreas and blood stream. There are several theories regarding the cause of the disturbed carbohydrate metabolism. It may be due to a sluggish or insensitive secretory response in the pancreas; perhaps the affected person has some defect in his tissues requiring unusual amounts of insulin for normal glucose metabolism; or insulin may be rapidly destroyed, inhibited, or inactivated in the diabetic person. Not all of these theories receive wide acceptance. The juvenile diabetic patient is very labile and can be controlled only with insulin injections and diet individualized to his needs. The more stable adult-onset diabetes can often be controlled with hypoglycemic agents (which stimulate insulin production by islet cells) given orally in conjunction with diet management.

HEREDITY AND INCIDENCE. Diabetes occurs in almost 2% of the total population but increases in frequency with advancing age. Estimates by the United States Public Health Service are: 2 in 1,000 persons under age 24; 10 in 1,000 in ages 25 to 44; 33 in 1,000 in the age group 45 to 54; 56 in 1,000 between ages 55 and 64; and 69 per 1,000 between ages 65 and 74.[5] The heritability of diabetes is unquestioned, but again the mechanism is unknown. Some consider diabetes to be due to an autosomal-recessive gene at a single locus with only a 20% penetrance; others consider it to be autosomal-dominant with incomplete penetrance. Another theory proposes a major gene with lesser modifying genes. It has been suggested that the juvenile-onset diabetic is homozygous for a specific gene while the maturity-onset diabetic is heterozygous. There is some doubt whether the adult and juvenile forms are two distinct diseases, and it is even postulated that diabetes may be a number of distinct disorders of carbohydrate metabolism, each due to a gene at different loci. However, most authorities favor a multifactorial inheritance. The importance of environmental factors, especially diet, cannot be overlooked. Exogenous factors may precipitate diabetes in genetically predisposed individuals; for example, pregnancy, certain diuretic drugs, and obesity are frequently associated with the disorder.[11]

RISK ESTIMATES.[14] The risk of diabetes in a family in which there is an affected member is derived from published data rather than any known genetic mechanism, and the age of onset is important. There is an unusually high incidence of early-onset diabetes among first-degree relatives. The concordance in monozygotic twins is 48% to 80% and the concordance in dizygotic twins is 3% to 5%, which indicates a decided genetic predisposition. Risk estimates vary, depending on the investigator, but all agree that the earlier the onset of diabetes in the index case the higher the risk for sibs. The risk doubles if a parent also has early-onset diabetes but is less if the onset was in middle age. The risk of having an affected child is 1 in 10 if one parent has early-onset diabetes and 1 in 5 if both parents have the early-onset disease; the risk of an affected child is 1 in 20 if the parent has adult-onset diabetes. If one parent has late-onset and one has early-onset diabetes, the risk is probably somewhere intermediate. It appears that genetic factors are important in early-onset diabetes and environmental factors are more significant in the late-onset disease.

Essential hypertension

High blood pressure, hypertension, is sometimes secondary to pathology in other tissues or organs (for example, renal disease, endocrine abnormalities), is relatively uncommon, and has its onset in early adulthood. Another form for which no etiology can be identified first appears in middle age and is referred to as *essential hypertension*. It is generally agreed that there is a genetic component in essential hypertension, but the nature of the component has not been resolved. Some consider essential hypertension to be multifactorial, while others favor a single-gene etiology.

Support for multifactorial etiology is consistent with the findings that the frequency distribution of blood pressures at different ages is continuous, with no division between normal and abnormal. In other words, blood pressure results from the action of several genes, and hypertension merely represents one extreme of the normal distribution curve. After all, blood pressure is the result of many factors, such as pulse rate, diameter and elasticity of blood vessels, cardiac output, state of the vasomotor center, sensitivity of vessels to nervous stimulation, and so forth. There is a relationship between blood pressures of close relatives and the proband that is consistent with the regression in the normal distribution curve. The blood pressures of close relatives are lower than the affected person's but higher than the population mean. The *degree* of this relationship is the same regardless of the blood pressure of the proband.

Argument for a monofactorial (single-gene) etiology is based on tests made with middle-aged men (age 45 to 60) of the same occupation. When blood pressures of this group were recorded they were found to be continuously distributed; however, when this same group was classified according to longevity of their parents, a different distribution pattern emerged. The systolic blood pressure of men whose parents lived to 65 years of age or beyond showed the same continuous distribution; the systolic blood pressure values of those who had a parent who died in middle age (40 to 64) showed a bimodal distribution: one peak at 140 mm Hg and a second peak at 175 mm Hg. This finding is clearly indicative of a single gene with a dominant effect. It is also suggested that the high blood pressure represents the homozygote and a moderately high pressure represents the heterozygote.

The controversy is unresolved, but if increased blood pressure is due to a single gene it would follow that there is a biochemical defect responsible for the abnormality. The discovery of such a defect would greatly aid in diagnosis and treatment, although the secondary effects of environment should not be overlooked.

Peptic ulcer

Ulcers in the gastrointestinal tract are classified according to location. Gastric ulcers affect the lining of the stomach, while duodenal ulcers involve the portion of the small intestine leading from the stomach, the duodenum. In these areas the pH of the gastric contents have the highest acid content. There is evidence to indicate that heredity plays a role in the cause of both types of peptic ulcer because of increased frequency among relatives, concordance in twins, and relationship to blood group factors. Duodenal ulcers comprise about 80% of peptic ulcers and are most common in men between the ages of 20 and 50 years. More males are affected than females, but this difference is not evident before puberty and in older age groups. Gastric ulcers occur less frequently and are relatively more common in females. Both gastric and duodenal ulcers are twice as common among first-degree relatives of affected persons, and the type of ulcer tends to be the same in both the proband and the affected relative. The concordance in monozygotic twins appears to be twice as high as in dizygotic twins, but not so high as to eliminate environmental influences as important factors in etiology.

Persons with group O blood are 35% to 40% more susceptible to duodenal ulcer (and to a lesser extent gastric ulcer) than those with other blood groups. Also ulcers, especially duodenal, are 50% more common in persons who are non-secretors of ABO substances (p. 120) than in secretors. It is postulated that the integrity of the duodenal and gastric mucosa are under the control of several genes predisposing to ulceration. This may be due to the decrease of blood group antigens on the mucosal lining in nonsecretors, which would implicate an immunologic factor. Also, group O is associated with a larger gastric acid output and secretory mass than is group A. It has been found that nonsecretors may have less muco-polysaccharide protecting the mucous membrane of the stomach. Individuals who are both type O and nonsecretors are about 25 times more prone to develop ulcers than the least susceptible persons who are secretors of A, B, or AB antigens. There is some question regarding the relationship between female hormones and the decreased incidence of peptic ulcers in women between puberty and menopause.

The relative importance of the genetic or environmental factors is unclear. Peptic ulcers do tend to run in families, and each type tends to appear in different families. Environmental stresses are important in etiology, as evidenced by an increased incidence in males in the age group most liable to be subject to stress and worry. Ulcers are known to be induced by stresses such as severe burns and the administration of stress hormones, for example, corticosteroids. Also, different environmental influences probably account for an increased incidence of duodenal ulcers in executive classes, whereas gastric ulcers are more common in lower social classes.

Ischemic heart disease[3, 14]

Ischemic heart disease (IHD), sometimes called *arteriosclerotic heart disease* or *coronary artery disease,* is well known for its contribution to mortality statistics. The risk of death from IHD in the general population is estimated at about 1.5 in 100 for males and 1 in 100 for females. The extent of genetic influences in IHD has been of interest for some time. Little is known about the relationship between genetic factors and development of IHD, but there is an increased risk in persons with hyperlipemia, diabetes, and hypertension, all known to be influenced by heritable factors. There is considerable familial incidence, and it is particularly evident where index cases are relatively young (under 55 years of age in males and 65 in females) and free from hypertension. One study that demonstrated increased evidence of IHD in first-degree relatives showed that when the index case was male, 1/12 of male relatives and 1/36 of female relatives were affected; when the index case was female, 1/10 of male and 1/12 of female relatives are affected.

The role of environment is evident in many instances. The incidence of IHD is higher in technologically advanced countries. It is uncertain whether this is related to specific dietary factors (high in unsaturated fats and refined carbohydrates) or other factors such as smoking, obesity, and lack of exercise. IHD is rare among underdeveloped countries; however, when persons from such backgrounds adopt "civilized" habits, IHD rapidly appears among them. For example, IHD is exceedingly rare in some African populations, but no distinction is apparent in Africans in Johannesburg or African descendents in the United States.

Epilepsy

There are few diseases that generate as much fear and anxiety among relatives as epilepsy. For many it represents the archetype of severe hereditary affliction. Epilepsy or *recurrent convulsive disorder* is the term used to designate a variable symptom complex characterized by one or more of the following: recurrent loss of consciousness or impaired consciousness, excess muscle movement, abnormal behavior, or sensory disturbances. Classification is made on the basis of etiology (genetic or acquired), the clinical features (type of seizures or behavior), the area of the brain involved, or electroencephalographic (EEG) patterns. If a brain abnormality can be demonstrated, the epilepsy is considered to be *organic* or *symptomatic* of a disorder; in the absence of a causative factor, the disease is said to be *idiopathic* (self-originating or of unknown cause) or *cryptogenic* (of hidden or obscure origin).

It is probable that some genetic defect in cerebral metabolism may be responsible for many cases of idiopathic epilepsy. It is known that persons predisposed to epilepsy have seizures when their basal level of neuronal excitability exceeds a critical point or threshold; no attack occurs if the excitability is maintained below this threshold. The administration of anticonvulsive drugs serves to raise this threshold and prevent seizures. It is possible that seizure activity is the result of excessive excitability of the neurons or deficiency in the suppressive influence of an inhibitory mechanism. In the development of such a complex structure as the brain any minor deviation that might take place could be expressed in abnormal behavior, one form of which is seizures.

Seizures are a feature of many neurologic and metabolic disorders and diseases with central nervous system involvement, including heritable diseases such as phenylketonuria or Tay-Sachs disease. Based on characteristic activity and EEG tracings four major subdivisions are recognized:

1. *Grand mal* is a seizure in which the individual loses consciousness and there is abnormal muscle action. The form of muscle activity is either *tonic* (persistent contraction of the muscles) or *clonic* (alternating contracting and relaxing of muscles). The convulsions may be *generalized,* reflecting overall electrical disturbance; *focal,* indicating a disturbance confined to a specific localized area; or *Jacksonian,* when the disturbance begins in a focal area and then spreads to other areas of the brain and is manifested by muscle contractions in one part of the body that progressively "march" to adjacent areas.

2. In *petit mal* seizures there is generalized brain dysrhythmia but the discharge is cut short before a convulsion develops, producing relatively minor manifestations. These are *myoclonic,* in which a burst of neuronal activity results in a single violent muscular jerk, and *absent,* characterized by a

transient loss of consciousness with minor manifestations described as "dizzy spells," "lapses," or "staring."

3. *Psychomotor epilepsy* is evidenced by purposeful but in appropriate behavior that may be repetitive and often complicated.

4. *Autonomic* epilepsy is manifested by physical symptoms such as pallor, rapid heart rate, flushing, perspiration, or other visceral symptoms.

Idiopathic epilepsy is difficult to define, and the genetic component is not clearly demonstrated. Epilepsy and disturbances in the EEG have been observed in identical twins, and the incidence of seizures in relatives of affected persons is two or three times higher than in the general population. Although seizures are known to accompany many forms of mental retardation, there is ample evidence to indicate that persons with epilepsy may be of high intelligence: well-known examples are Julius Caesar, Alexander the Great, Socrates, Tchaikovsky, Napoleon, van Gogh, and Tolstoy.

Children who have had some convulsions in childhood are more liable to develop epilepsy than children who have never had a convulsion. There is also an increased frequency of childhood convulsions and epilepsy in sibs of these children. The risk is higher if the convulsions occurred between the ages of 1 and 5 and is increased if there is a positive family history in either parent. Abnormal EEG tracings are found more often in parents and siblings of affected persons than in the general population. It is estimated that if a child in a family has epilepsy, subsequent children have an 8% to 13% chance of being similarly affected. If a parent has the disorder, there is a 3% to 5% risk for seizures in an offspring. The risks vary in relation to the type of seizure, but about 30% of first-degree relatives will show characteristic EEG patterns some time during their life. In general, the incidence in relatives is low. There is one rare variety that is due to a dominant gene with a high degree of penetrance and seems to be typically of late onset.

A number of environmental factors, such as drugs, overhydration, alkalosis, and photic or other sensory stimuli, can precipitate seizure activity in susceptible persons. In some varieties of epilepsy abnormal EEG patterns follow interrupted light stimulation although not all will respond by convulsions. However, abnormal EEG response to light is high in some families, often as high as 50% in first-degree relatives.

Infectious diseases

It appears that the genotype of a person can have a decided effect upon the susceptibility and clinical course of some infectious diseases.

Tuberculosis. Tuberculosis is an infection caused by the *Mycobacterium tuberculosis*. The organisms are strictly aerobes (those that grow in the presence of free oxygen) and therefore thrive best in tissue with a high oxygen and low carbon dioxide content. The organs most often affected first by the tuberculosis bacillus are the apex of the lungs, the kidneys, and the growing ends of long bones. The liver and spleen, with low oxygen tension, are rarely affected. Organisms are usually transferred by close indoor contact of a susceptible person with an affected individual who is excreting the bacillus from his lungs in large numbers by coughing, sneezing, and talking.

Although the clinical course of tuberculosis is related to age, sex, and living conditions of affected persons, the resistance to the disease appears to be modified by the genotype. Some racial groups demonstrate a pronounced vulnerability to tuberculosis: American Indians, Eskimos, and Negroes. There is a significantly higher degree of concordance between monozygotic twins (74%) than dizygotic twins (28%). This marked difference between the two types of twins indicates that infection alone is not enough to produce disease. Development of the disease is dependent upon the genotype approximately 46% to 76% of the time and to environmental influences 26% to 54%. It appears that the clinical course and outcome of the disease depend relatively little upon genetic factors, but the susceptibility upon exposure to the bacillus is decidedly affected by the genetic constitution.

Paralytic poliomyelitis. It has been found that paralytic poliomyelitis is 36% concordant in monozygotic twins but only 6% in dizygotic twins. It appears that the disease does not depend entirely upon the infecting organism that most human beings are normally exposed to. The incidence of the disease is dependent upon hereditary factors about 35% of the time and environmental factors about 65%.

Malaria. The most well-documented example of resistance to infection is provided by the sickle cell trait (p. 64). Populations that have been exposed to the *Plasmodium falciparum* malaria have a high incidence of the trait. In Congolese populations, where malaria is prevalent, the parasite counts in the blood of persons with HbAS is significantly lower than in persons with normal hemoglobin. The presence of the sickling hemoglobin confers a natural resistance to the malaria. The same advantage seems to be bestowed on persons with the minor thalassemias (p. 66).

Mental deficit

Intelligence, like stature, shows a continuous variation. Distribution in the general population extends from retardation (IQ less than 70) to exceptional (IQ greater than 130), with the bulk around a mean of 100 (50% with IQ between 90 and 110). It is estimated that approximately 75% to 85% of intelligence is contributed by the small additive effects of polygenes and the remainder by environmental factors such as stimulation. Severe mental deficit is represented by persons with an IQ of less than 50 and is caused by inborn errors of metabolism, chromosomal aberrations, and congenital malformations, as well as infection, trauma, hypoxia, and so on. Mild retardation, represented by persons with an IQ around 70, is considered to be at the lower end of the normal distribution curve and is probably the result of both polygenic inheritance and cultural deprivation. There is little difference in sex proportions in intelligence level down to IQ of 55; below this level the incidence in males rises steadily, probably indicating an X-linked recessive gene effect.

The correlation between sibs is high in the group with IQ over 50 but not in the group below this point, which reflects the influence of other factors in severe mental deficit. The correlation between intelligence of identical twins is very high: 87% in twins reared together and 75% in those reared apart. The difference is greater in nonidentical twins (55% to 65%). Close relatives of mildly retarded persons demonstrate the usual regression toward the mean of the population as a whole. Their intelligence is lower than the population mean but higher than that of the retarded individuals. Risks to first-degree relatives is the same; that is, the probability is that the offspring of mildly retarded persons will produce children with intelligence approximately midway between their own and the population mean. Persons in this group are usually of average or increased fertility and with few exceptions mate with others near their own level of intellect.

First-degree relatives of severely retarded persons show a different pattern. If the intellectual deficit is due to environmental factors such as injury or infections, the relatives will not be affected. Where the retardation is due to a mutant gene the probability that relatives will have the same abnormality is based on the risks related to the inheritance pattern of the gene. The same is true if the deficit is due to a chromosomal aberration. Severely retarded persons are almost always infertile.

Mental illness

Schizophrenia.[12] Early in the century Eugen Bleuler coined the word schizophrenia (*schizein,* to divide; *phren,* mind) to describe his interpretation of a mental disorder in which there is a cleavage or disunity of the mental functions. The disorder is characterized by an alteration in the manner in which the affected individual thinks, feels, and relates to the external world. The onset of symptoms usually occurs in adolescence or early adult life and may be dominated by one of several symptom complexes that are not integrated with drives or ideas. The four A's of schizophrenia are disturbances in *affect* (outward expression of emotion), loss of continuity of thought *associations, autism* (living in one's own inner world, detached from reality), and *ambivalence* (having opposite or contrary feelings, desires, or thoughts simultaneously). Fragments of ideas are connected in illogical ways; emotions may be lacking or in disagreement with intellectual processes; there are bizarre, unpredictable, or disconnected thought associations or they may be lumped together; and concepts lose their completeness. There is progressive deterioration of the personality.

Secondary symptoms, or those that express the disorders of thinking, emotion, and attention include hallucinations (perception of the senses not found in objective reality, particularly hearing of nonexistent voices); delusions (false beliefs unable to be substantiated by evidence), especially of persecution; confusion, stupor, feelings that fluctuate between excitement and melancholia; and catatonia, peculiar motor signs such as immobility, rigidity, mutism, and stereotyped posture and activity.

Four major subtypes are: *paranoia,* in which delusions and hallucinations are the predominant symptoms; *catatonia,* marked by the catatonic motor signs; *hebephrenia,* usually an acute early-onset type characterized by shallow, inappropriate affect, with silly behavior and mannerisms; and *simple,* marked by progressive apathy and indifference but in which secondary signs do not appear.

There is considerable disagreement concerning the cause of schizophrenia. Some favor a biochemical basis for the disorder, while others support the theory of a complex combination of psychosocial and environmental stresses. There is ample evidence to indicate that genetic factors contribute appreciably to its development; this discussion will be limited to the conclusions supporting this concept.

The incidence of schizophrenia is estimated to be about 1% of the population, with no racial or sex differences. Most investigators indicate a very high frequency in families: schizophrenia occurs in 11% of siblings of affected patients. The concordance rate for monozygotic twins in which one member is affected is 40% to 60% as opposed to 10% to 14% in dizygotic twins. This concordance was maintained even in twins who were reared in separate environments, and these twins developed the schizoid symptoms at about the same age. Comparisons of groups of children adopted before one month of age (to minimize the effect of an affected mother) showed that 10% of those whose biologic mother was schizophrenic developed the disorder, while none of those born to nonschizophrenic mothers became schizoid. There is considerable variation among studies, but the evidence for a genetic factor in the contribution to the development of schizophrenia is strong and found consistently. When one parent is affected the estimated risk to offspring is 1 in 8 to 12; if both parents are affected, the risk is 2 in 5. The risk to siblings in a family with an isolated case is probably 1 in 8 to 15. There is a clear correlation between the degree of blood relatives and the incidence of schizophrenia.

Several modes of inheritance have been advanced for schizophrenia. Both autosomal-recessive and an intermediate gene, manifest in both homozygote and heterozygote, have been proposed, but neither of these is sufficiently supported by data. The most probable explanation seems to be a polygenic predisposition that interacts with as yet unspecified environmental factors to produce the manifesting phenotype or an autosomal-dominant gene (with a reduced penetrance) with multiple gene modifiers and environmental factors. From accumulated data it has been estimated that the heritable component probably comprises about 73% when environmental conditions are uncontrolled or unspecified.

Some inherited characteristics related to schizophrenia may throw some light on its etiology. Using a variety of psychologic tests, one group of investigators found a disorder of concept formation in relatives of patients with schizophrenia. Others have found a significant difference in the total finger ridge counts in analysis of fingerprints of affected persons compared with normal persons (Appendix C). The difference has important implications since the ridge count is almost entirely genetically determined.

It has been known that there is a biochemical difference between normal individuals and those with schizophrenia. A very high percentage of schizophrenics have a lactic acid–pyruvic acid (L/P) ratio greater than normal, and recently it has been found that there is also an elevated L/P ratio in some relatives of schizophrenics. Another interesting hypothesis regarding a biochemical basis in schizophrenia is that in affected persons there is a metabolic abnormality that causes the conversion of some physiologic substances into hallucinogenic ones. Both findings need further investigation.

Affective psychosis.[12] Affective psychosis, or *manic-depressive psychosis,* is a mental illness with a variable course. An affected individual may have one attack or many; the attacks are self-limiting but may be short or long in duration; and the manifestations can be primarily manic or depressive or can alternate between the two. During attacks of mania the affected person feels a sense of elation manifested by mild symptoms of increased activity, talkativeness, and random thoughts; or there may be a severe attack, with extreme activity and incoherence among the symptoms. The depressive states may range from listlessness, indifference, low energy level, and disturbed sleep, appetite, and concentration to feelings of worthlessness, tearfulness, or self-abnegation, with constant pacings and hand-wringing behavior. Affected persons are fully oriented although some delusion appropriate to the affect may be manifest, and between attacks they appear to be quite normal.

As in schizophrenia, the interaction of multiple components make the understanding of genetic and environmental contributions to the affective psychoses difficult to understand. The incidence in the population seems to be about 1%, and frequency in first-degree relatives is higher than would be expected by chance. Concordance for the disorder is 60% in monozygotic twins and 25% in dizygotic twins, and the frequency in sibs of affected persons is 10%. The evidence reflects a strong genetic influence, but the mode of inheritance is unknown. Many theories have been proposed, including various patterns of dominant, recessive, and X-linked genes acting individually or in combination. Many favor a multifactorial influence. The proposed patterns are much the same as those proposed for schizophrenia. There does seem to be an increased risk to offspring and sibs if the proband has episodes with mania or mania and depression than where there has been depression alone.

CANCER[7, 15]

Cancer, a malignant tumor or malignant neoplasm (*neo,* new; *plasm,* formation), is the proliferation of new cells that are characterized by uncoordinated growth with no sharply delineated borders. As growth progresses, the cells become increasingly disorganized and take on an atypical morphology and function. One of the most characteristic cellular changes is a tendency toward loss of a differentiated, specific function to become less specialized, much like their undifferentiated precursors. This undifferentiation is known as *anaplasia.* It is displayed to some degree in all malignant neoplasms and is one of the most reliable indications of malignancy. In general, the more anaplastic the cells, the more malignant is the neoplasm. The cells, inconsistent in size and shape, become increasingly autonomous. As they multiply, cancer cells progressively infiltrate, invade, and destroy adjacent tissues and disseminate to remote sites, producing secondary growths, or *metastases.*

The mechanism by which malignant cells are produced is still unclear although much has been learned. It is known that certain environmental agents that are mutagenic (for example, ionizing radiation and chemicals) are also carcinogenic; that is, they are capable of inducing cancer in experimental animals and have been implicated in the production of cancer in man. A mutation that changes the genetic material will be transferred to all descendents of the altered cell. Constant irritation may be a factor, and viruses have produced such alterations in experimental animals. Loss or failure of the immune surveillance may play a role. It is felt that altered cells contain antigens that stimulate the immune response just as occurs with antigens of invading organisms; if the immune mechanism does not destroy these cells they are free to proliferate. Another plausible theory concerns loss of repressor control following a somatic mutation; the cell fails to respond to the normal regulatory mechanisms.

Cancer is very common, affecting 1 in every 4 persons at some period in their lifetime, and accounts for 15% to 20% of mortality from all causes. The frequency is highest at the extremes of the life span, with 50% occurring after age 65. Tumors of childhood appear to arise in tissues that are undergoing rapid growth or in primitive cells probably related to organogenesis. In the older age groups cancer is related to prolonged exposure to carcinogens or to organs undergoing involutionary changes (such as prostate or uterus) and those with rapid turnover of cells (such as the lining of the gastrointestinal tract).

The role of heredity in the etiology of cancer is a relatively minor one. Concordance rate in twins shows no difference in monozygotic and dizygotic twins for total incidence of cancer; but when related to cancer of a specific organ, monozygotic twins show a higher concordance rate than do dizygotic twins. In some races there is some variation in the incidence and type of cancers, but it is uncertain whether this is due to racial differences or the environment. For instance, breast cancer is common in the United States and Europe but uncommon in the Orient; cancer of the liver is common in Central and South Africa but rare in Europe and North America; facial skin cancer is common in light-skinned persons exposed to large amounts of sunlight in tropical regions, while in Negroes the incidence of facial skin cancer is very low.

There is no general heritable predisposition to cancer, but there is evidence to indicate that within some families there is an inherited predisposition to certain forms of cancer. In most specific-site cancers the risk to first-degree relatives of a person with cancer is three times that of the population as a whole. As mentioned previously, persons with blood group A are more liable to develop cancer of the stomach, and rare families have been reported in which the incidence is extremely high although the explanation is not clear. There are a few rare precancerous conditions with mendelian inheritance patterns, and there appears to be a higher incidence of cancer related to some chromosome abnormalities.

Heritable predisposition to malignancy

A representative example of some disorders with a genetically determined susceptibility to malignant disease will be briefly described.

Retinoblastoma. This relatively rare malignant tumor of the retina is inherited as autosomal-dominant, although most cases appear to be fresh mutations. The incidence of the tumor is approximately 1 in 20,000 and is seen almost exclusively in children less than 5 years of age. The tumor arises in the retina and if untreated invades the optic nerve, then the brain. It can be unilateral or bilateral; 25% are bilateral and can appear simultaneously or successively. If detected early, light coagulation, x-ray treatment, and chemotherapy are preferred over enucleation as treatment. Retinoblastoma is an example of a disorder that demonstrates reduced penetrance. It has been estimated that 10% of children who have the dominant gene do not develop tumors. There are some who feel that unilateral cases represent somatic mutations and that only bilateral cases are transmissible. Since there is a 50% risk that children of affected parents will develop this type of tumor, early and frequent ophthalmic examinations are carried out on all children of an affected parent.

Neurofibromatosis. This cancer, sometimes known as *Recklinghausen's disease,* is transmitted by an autosomal-dominant mode of inheritance and is highly variable in its expression. Some affected individuals exhibit numerous fibrous tumors along peripheral nerves that vary in size from tiny nodules to large masses; others have only flat pigmented areas of skin, "cafe au lait spots," also varying in size and number. These tumors or discolored areas (5% to 8%) may undergo change to neurofibrosarcoma, cancer of the nerve sheath. There is no relationship between the type and extent of cutaneous manifestations and the development of cancer. Some persons with extensive tumor masses remain cancer-free, while persons with relatively minor skin pigmentation develop malignancy.

Multiple polyposis of the colon. This condition is clearly due to a dominant gene. The disorder is manifest by many polyps (smooth projecting growths) extending throughout the colon and rectum. These diffuse polyps frequently give symptoms of diarrhea or bleeding, and there is an exceedingly high incidence of cancer of the colon in affected persons. The probability is that without surgical treatment affected persons will develop cancer eventually. The treatment is surgical removal of the colon.

Xeroderma pigmentosum. This precancerous disorder of the skin is inherited as an autosomal-recessive trait. There are multiple manifestations that may include neurologic symptoms, ocular disturbances, and some metabolic abnormalities. Cancer changes arise on the atrophied keratotic tissue of the skin and are frequently precipitated by exposure to sunlight.

Cytogenetics and cancer

Practically all cells of normal persons demonstrate normal karyotypes. Cells from virtually all solid cancers and many blood tumors show abnormal karyotypes. Mitosis is often abnormal with various abnormalities in structure and number of chromosomes, particularly noticeable as cancer progresses. There are some associations between chromosome abnormalities and malignancies of the blood and blood-forming tissues.

Leukemia. The risk of developing leukemia is 20 times greater in persons with Down's syndrome (p. 80) than in the general population, although the overall incidence is only 1 in 500 persons with Down's syndrome. There have also been reports that Down's syndrome occurs more frequently in siblings of leukemic children who do not have Down's syndrome. These findings suggest a related mechanism.

Leukemia is associated with some disorders with mendelian modes of transmission. *Fanconi's aplastic anemia* (a deficiency of all cellular elements of the blood) and *Bloom's syndrome* (dwarfism and skin changes) both have chromosome anomalies and a predisposition to leukemia. Persons with the X-linked recessive traits of Wiskott-Aldrich disease (p. 134) and agammaglobulinemia demonstrate a higher incidence of leukemia.

There is one cancer, *chronic granulocytic leukemia,* in which there is a consistent recognizable cytogenetic abnormality. In the majority of cases one of the G group chromosomes (p. 9) has a partial deletion of the long arms and has been labeled the Philadelphia (Ph[1]) chromosome. It is unique to and a fairly constant feature of the disease and may represent the initial change that results in the leukemia.

Environmental agents known to produce chromosome breaks are irradiation and some chemicals. The effect of exposure to irradiation is well known, and an increased risk of developing leukemia has been shown in some survivors of the atomic bomb dropped on Hiroshima. Also, chemicals such as LSD may cause chromosome damage in users and possibly in children born to mothers who took LSD while pregnant.

PHARMACOGENETICS

Drug sensitivity of varying degrees is very common; some people are sensitive to the effects of a given drug, while others are resistant. There are many factors that affect an individual's response to drugs (such as age and concurrent diseases). The term pharmacogenetics applies to that branch of genetics concerned with drug responses and their genetic modification. In its broadest sense pharmacogenetics can be said to include any genetically determined variation in drug response. In a narrower view, pharmacogenetics is restricted to genetic variations that are revealed *only* by response to drugs. Under ordinary circumstances such gene traits are relatively harmless and the carrier is without symptoms, but in these persons a specific drug may cause severe or even fatal reactions. Genetically determined drug sensitivity is a good example of the way in which a trait can be uncovered only by an environmental stimulus: although the defect is present from birth, the appropriate environmental agent is required to reveal its presence.

Some diseases are precipitated by drugs; for example, porphyria (p. 55) in response to barbiturates, diabetes mellitus (p. 150) after administration of steroids, and systemic lupus erythematosus (p. 131) with administration of the antidepressant drug hydralazine. Others were discovered by the use of drugs: glucose-6-phosphate dehydrogenase deficiency (p. 52), in which affected individuals develop a hemolytic crisis when they take primaquine, aspirin, or sulfonamides; inherited resistance to coumarin anticoagulants, in which massive doses are required in order to achieve a therapeutic effect in affected individuals with thrombotic conditions such as myocardial infarction; and actalasemia, a condition due to lack of an enzyme, in which the affected persons do not show the normal frothing response to hydrogen peroxide.

Taste sensitivity to PTC

Although the ability or inability to taste the compound phenylthiourea (phenyl-thiocarbamide, or PTC) is not a disease, it is a classic illustration of an inherited difference in the response to a drug. The ability or inability to taste PTC, which is decidedly bitter, is due to a pair of allelic genes. About 70% of the population possesses one or both dominant genes and can taste PTC; 30%, homozygous for the recessive genes, are nontasters. However, among the tasters there are two sub-groups: those who can taste PTC in very weak solutions and those who can taste it only in high concentrations. These variations are felt to be due to modifying genes. There is also a difference in the ability to taste among races. Nontasters are rare in Negroes and Chinese and they are almost unheard of in American Indians and Eskimos.

Isoniazid metabolism

Although it does not precipitate a disease, the rate of metabolism of isoniazid (INH), an important drug for treatment of tuberculosis, is affected by gene action. After isoniazid is ingested it is absorbed from the gastrointestinal tract into the blood stream, distributed to the site of cellular interaction, then broken down (in-activated) and excreted. In some persons the rate of inactivation and excretion, re-flected by the reduction in blood levels, is slower than it is in others. Apparently a dominant allele is responsible for the production of a liver enzyme that is active in the metabolism of the drug. Persons homozygous for the recessive gene do not produce the enzyme and therefore metabolize the drug very slowly. Rapid inac-tivators are normal homozygotes or heterozygotes. The phenotype has no influence upon the way patients with tuberculosis respond to treatment; however, slow in-activators are more likely than rapid inactivators to develop the side effects of the drug. This is an important consideration in medical management of tuberculosis.

There are significant differences between persons of various racial backgrounds in inactivation of INH. Approximately 50% of Caucasians and Negroes inactivate INH rapidly; as high as 90% of Eskimos and Japanese are rapid inactivators; and about 65% of Latin Americans inactivate INH rapidly.

Other drug side effects

There are other abnormal responses to drugs attributed to a single gene, such as a reduced ability to eliminate curare (a valuable adjunct to surgical anesthesia), which produces prolonged muscle paralysis during anesthesia in affected persons (autosomal-recessive). Also, serious side effects from chlorothiazide (a frequently used diuretic) can occur in persons with gout or diabetes. With the rapid pro-liferation of drug varieties it is expected that more and more inherited abnor-malities in drug response will be recognized. Most investigators feel that the major advances in and the potential of pharmacogenetics merit its consideration as a separate branch of genetics.

REFERENCES

1. Campbell, M.: In Watson, H., editor: Paediatric Cardiology, London, 1968, Lloyd-Luke, Ltd., Chapter 5.
2. Carter, C. O.: Inheritance of congenital pyloric stenosis, Br. Med. Bull. **17:**251, 1961.
3. Carter, C. O.: An ABC of medical genetics, Boston, 1969, Little, Brown and Co., Chapter 6.
4. Carter, C. O.: Genetics of common disorders, Br. Med. Bull. **25:**52, 1969.
5. Diabetes source book, No. 1168, Washington, D. C., May 1969, U. S. Department of Health, Education, and Welfare.
6. Fraser, F. C.: The genetics of cleft lip and palate, Am. J. Hum. Genet. **22:**336, 1970.
7. Lynch, H. T.: Genetic factors in carcinoma, Med. Clin. North Am. **53:**923, 1969.
8. McKusick, V. A.: Human genetics, ed. 3, Englewood Cliffs, N. J., 1969, Prentice-Hall, Inc.
9. Perloff, J. K.: The clinical recognition of congenital heart disease, Philadelphia, 1970, W. B. Saunders Co.
10. Porter, I. H.: Heredity and disease, New York, 1968, The Blakiston Division, McGraw-Hill Book Co.
11. Rimoin, D. L.: Inheritance in diabetes mellitus, Med. Clin. North Am. **55:**807, 1971.
12. Rosenthal, D.: Genetic theory and abnormal behavior, New York, 1970, McGraw-Hill Book Co.
13. Stanbury, J. B., Wyngaarden, J. B., and Fredrickson, D. S., editors: The metabolic basis of inherited disease, ed. 3, New York, 1972, McGraw-Hill Book Co.
14. Stevenson, A. C., Davison, B. C. C., and Oakes, M. W.: Genetic counseling, Philadelphia, 1970, J. B. Lippincott Co.
15. Walker, S.: Chromosomes and cancer. In Clark, C. A., editor: Selected topics in medical genetics, London, 1969, Oxford University Press, Inc.
16. Warkany, J.: Congenital malformations, Chicago, 1971, Year Book Medical Publishers, Inc.

GENERAL REFERENCES

Erickson, M., Catz, C. S., and Yaffe, S. J.: Drugs and pregnancy, Clin. Obstet. Gynecol. **16:**184, 1973.

Hay, S., and Wehring, D. A.: Congenital malformations in twins, Am. J. Hum. Genet. **22:**662, 1970.

LaDu, B. N.: Pharmacogenetics, Med. Clin. North Am. **53:**839, 1969.

Platt, R.: Logic and hypertension, Lancet **1:**899, 1963.

Price, J.: Genetics and schizophrenia. In Clark, C. A., editor: Selected topics in medical genetics, London, 1969, Oxford University Press, Inc.

Robbins, S. L., and Angell, M.: Basic pathology, Philadelphia, 1971, W. B. Saunders Co.

Roberts, J. A. F.: An introduction to medical genetics, ed. 5, London, 1970, Oxford University Press, Inc.

Thompson, J. S., and Thompson, M. W.: Genetics in medicine, Philadelphia, 1966, W. B. Saunders Co.

Winik, M., Brasel, J., and Velasco, E. G.: Effects of prenatal nutrition upon pregnancy risk, Clin. Obstet. Gynecol. **16:**184, 1973.

Wintrobe, M. M., and others, editors: Harrison's principles of internal medicine, ed. 6, New York, 1970, Blakiston Division, McGraw-Hill Book Co.

8

Genetic equilibrium

The basic principles of heredity that are applied to individuals and families can be applied to populations as well. The study of the distribution and behavior of genes in individuals of the same species constitutes a most important aspect of inheritance and variation—population genetics. Those in this specialized branch of genetics are concerned with the types and frequences of genes in populations and the forces that operate to initiate, maintain, and spread these genetic diversities. Genes exist within individuals, but the fate of their genes is strongly dependent upon factors concerning the population as a whole. Conditions that tend to alter the gene frequencies within a population include such factors as the size of the population, migration, mating regulations, fitness, mutations, and degree of isolation. The modification of gene frequencies over countless generations is the substance of evolution.

GENE FREQUENCIES IN POPULATIONS

A population is defined as an aggregate of similarly adapted, sexually interbreeding or potentially interbreeding individuals. The size of a population may vary from very large racial groups to small isolated communities. Usually a genetic population is considered to be a local group, or *deme,* in which each member has an equal opportunity to mate with any other member of the opposite sex within the group. All the genetic information or sum total of genes distributed among the reproductive gametes of an interbreeding group collectively form a *gene pool,* and the gene frequencies are the proportions of the various alleles of a gene within the population. The kinds and frequencies of any genotype in a population depend on the kinds of genes and their frequencies donated by parents of the preceding generation, much the same as the relationship between parents and their offspring. The smaller the population, the fewer the number of genotypes produced. In the absence of modifying conditions this frequency distribution of genes within a population will remain constant; that is, in a random-mating population, a specific pair of allelic genes tends to exist in the same proportion generation after generation. Just as single genes are inherited in families, gene frequencies are inherited in populations.

The Hardy-Weinberg principle

The basis of population genetics is the *Hardy-Weinberg Law,* independently proposed by a British physician, G. H. Hardy, and a German physician, W. Weinberg. A principle derived from Mendel's law of segregation (p. 25), it states that the proportion of genotypes in respect to a single locus will remain constant from generation to generation as long as mating is random; that is, there is a predictable frequency of heterozygotes and homozygotes present in the population at any one time. A gene pool can be thought of as a beanbag containing black and white beans, the black beans representing the dominant allele and the white beans representing the recessive allele. (The beanbag analogy is so common that population genetics is sometimes called beanbag genetics.) If there are only these two possible alleles, the three combinations that can be made from the black and white beans are two black, two white, and one black and one white. The chance of a person drawing any of these three combinations is estimated from the frequency of each color of bean contained in the bag. Where black and white beans are in equal proportion, half black and half white $= 1$, or 100%. If p is the frequency of the black beans and q is the frequency of the white beans, then $p + q = 1$ since this is the total of the beans in the population. Now let the black beans represent T (ability to taste PTC) (p. 163) and the white beans represent t (the inability to taste PTC). The chance that an individual selected at random will have the TT (homozygous dominant) genotype is p^2; the chance that the individual will have the genotype tt (homozygous recessive) is q^2; and the chance that he will have the genotype Tt (heterozygous dominant) is $2pq$. The 2 in the $2pq$ means that there are two ways to form the heterozygote pq, as illustrated in the Punnett square in Fig. 8-1. Keep in mind that the various combinations formed by p and q repre-

		Male	
	Gametes	p (T)	q (t)
Female	p (T)	p^2 (TT)	pq (Tt)
	q (t)	pq (Tt)	q^2 (tt)

Fig. 8-1. A standard model of frequency distribution when half the genes in a population are dominant (T) and half are recessive (t). Summary: p^2 (TT) + $2pq$ (Tt) + q^2 (tt).

	Frequency of male genotypes		
	p^2 (TT)	$2pq$ (Tt)	q^2 (tt)
Frequency of female genotypes — p^2 (TT)	p^4	$2p^3q$	p^2q^2
$2pq$ (Tt)	$2p^3q$	$4p^2q^2$	$2pq^3$
q^2 (tt)	p^2q^2	$2pq^3$	q^4

Fig. 8-2. Types of combinations and their relative frequencies in a population containing TT, Tt, and tt genotypes.

Table 8-1. Frequencies of genotypes of offspring in a random-mating population

Matings		Offspring		
Type	Frequency	TT	Tt	tt
$TT \times TT$	p^4	p^4		
$TT \times Tt$	$4p^3q$	$2p^3q$	$2p^3q$	
$TT \times tt$	$2p^2q^2$		$2p^2q^2$	
$Tt \times Tt$	$4p^2q^2$	p^2q^2	$2p^2q^2$	p^2q^2
$Tt \times tt$	$4pq^3$		$2pq^3$	$2pq^3$
$tt \times tt$	q^4			q^4

Sum of TT: $p^4 + 2p^3q + p^2q^2 = p^2(p^2 + 2pq + q^2) = p^2$
Sum of Tt: $2p^3q + 4p^2q^2 + 2pq^3 = 2pq\ (p^2 + 2pq + q^2) = 2pq$
Sum of tt: $p^2q^2 + 2pq^3 + q^4 = q^2(p^2 + 2pq + q^2) = q^2$

sent frequencies of all possible matings rather than a single mating. In a condition where a given locus contains only two alleles, these three genotypes represent the total genetic composition for this locus in the population. From this is derived the Hardy-Weinberg equilibrium:

$$p^2 \text{ (for } TT) + 2pq \text{ (for } Tt) + q^2 \text{ (for } tt) = 1$$

A significant consequence of the Hardy-Weinberg law is that the *proportions* of genes do not change from generation to generation. The frequencies of offspring from these matings are represented in Fig. 8-2 and Table 8-1. Using all possible matings in each generation, the proportions of T and t remain the same as they were in previous generations. In each generation half the genes are T and half are *t*. Dominant genes do not replace recessive alleles in a stable population. Dominance and recessiveness are the expressions of the genotype, not the frequency of the genes. This explains why genotype frequencies for such traits as brown eyes and blue eyes, tasters and nontasters, and so on, are maintained in the same proportions in a population. In a random-mating population an equilibrium will be reached in a single generation and maintained thereafter.

Frequencies of rare genes

The Hardy-Weinberg formula produces the same consistent results when gene frequencies are not equal; that is, instead of half dominant and half recessive, the frequencies may be three-fourths dominant and one-fourth recessive. However, each generation will contain the same *proportion* of alleles as found in the previous generation. The gene frequencies are at equilibrium. The usefulness of this formula is that it provides a means to calculate an estimate of the frequency of heterozygotes in a population when the frequency of a rare homozygote (such as phenylketonuria or albinism) is known. For example, albinism (p. 51) is an autosomal-recessive disorder that occurs once in every 10,000 persons (q^2). The frequency of this recessive allele is, then, $\sqrt{1/10,000}$ or $1/100$. The frequency of p can be calculated using the formula $p + q = 1$: $p = 1 - 1/100$, or $99/100$. The frequency of heterozygous carriers of albinism ($2pq$) is, then, $2 \times 1/100 \times 99/100$, or approximately 1 in 50. In the general population, carriers are much more common. If the disorder is due to a dominant gene, heterozygotes are impossible to distinguish from homozygotes; therefore, the frequency of heterozygotes is calculated from the frequency of recessive homozygotes in the same way as the gene frequency was calculated for albinism.

FACTORS OF DISTURBANCE

When gene frequencies do not fit the Hardy-Weinberg formula it can be suspected that some factor is operating to disturb the equilibrium. Factors that most frequently disturb the Hardy-Weinberg law are nonrandom mating, genetic drift, migration, mutation, and selection.

Nonrandom mating

The Hardy-Weinberg equilibrium can be maintained only if there is random mating. In large populations, mating is usually random for most traits; in other words, the gene constitution of an individual usually has no influence on his choice of mate. Where there are individuals of the three genotypes who are equally likely to marry and produce offspring, persons of any one genotype may select spouses without demonstrating preference for any particular genotype. In fact, regarding traits such as the ABO blood groups, even if the type were known it would not influence the choice of a mate (other than as related to ethnic background) any more than would the color of a person's eyes. The technical term for the random association of genotypes is *panmixis*.

If an individual shows a preference for a person of a particular genotype it is referred to as *nonrandom*, or *assortative mating*. Assortative mating is not uncommon. Within any population there are numerous subgroups who differ genetically (for example, races with characteristic gene frequencies) or nongenetically (for example, religious preferences). Members of such groups are likely to intermarry. Within most populations persons tend to assort in regard to some traits: intelligence, stature, socioeconomic status, for instance. An example of assortative mating for a common physical defect is that between deaf mutes, who tend to aggregate for social purposes. The development of organizations designed to encourage the meeting of persons with similar genotypes may increase the frequency of assortative matings among persons with other disorders, such as diabetes, hemophilia, and so on.[2] Assortative mating is positive when persons resemble one another for a trait; when they show a preference for opposite types it is negative assortative mating. The primary consequence of assortative mating is an increase in the proportion of homozygotes in a population. When homozygote marries homozygote, the frequency of that gene increases in the population.

Consanguineous matings are sometimes considered a form of assortative mating. Relatives have more genes in common, including rare recessive genes, than persons in the population as a whole. Consanguineous matings do not change the gene frequencies in a population; however, they do change the genotype frequencies by increasing the proportion of homozygotes.

Genetic drift

Genetic drift is a term used to describe the extinction or fixation (frequency of 100%) of a specific gene that occurs by chance in small isolated populations. Isolates, small groups that form for geographic, political, religious, or social reasons, are not common at the present time, but they probably contributed to variations in gene frequencies in the past when the population was smaller and man lived in small isolated groups. Drift undoubtedly occurred in subgroups that migrated to start new communities, particularly since gene frequencies in small migration groups probably did not accurately reflect the gene frequency of the population from which it was derived. To illustrate genetic drift, assume that in a small isolated population recessive gene *a* is less common than dominant gene *A,* which is possessed by nearly everyone. If the few persons with gene *a* either do not have offspring or pass only the *A* gene to their offspring, it is possible that the *a* gene could disappear from the population in a generation. In this situation the *a* gene has become extinct and the *A* gene has become fixed in the population.

A frequently used example of genetic drift is the blood group composition of American Indians. There is a high frequency of blood group O in most tribes, yet in the Blackfoot the blood group is primarily type A. In this tribe the recessive homozygote (assuming all American Indians were originally of the same genotype) has been almost eliminated. Also, a small religious isolate in Pennsylvania, the Dunkers, were found to have blood group frequencies quite different from the parent population in Germany.[3]

A present-day example of genetic drift is the founder effect demonstrated by McKusick's studies of the high frequency of a defect found among the Old Order Amish, a social isolate group in Lancaster County, Pennsylvania. A recessive gene in the homozygous state produces the rare Ellis–van Creveld syndrome that includes a particular form of drawfism accompanied by an extra finger. Extremely rare in the population as a whole, the disorder is found to be as high as 1 in 160 among this isolated group, with an estimated heterozygote frequency of 13%. Apparently one of the founders was a carrier of the gene, and random genetic drift contributed to its increase in frequency. Amish populations living in other locations with different founders do not have this abnormality.[5]

Gene flow

A variation in gene frequency in small populations, genetic flow, or migration, is a slow, gradual change in the gene frequency in larger populations as a result of the addition of new genes. When two widely different groups continually interchange members, the populations slowly become more similar in gene frequencies. The classic example of gene flow is the steady cline (a geographic gradient in gene frequency) from a high blood group B and low group A in Asia to a low blood group B and high A in Europe. The interchange of genes among groups within close proximity to each other gradually alters the gene constitution in the direction of the flow.

The forced migration of African peoples to the United States has produced a change in the gene frequencies measured by blood group alleles. The flow has been almost exclusively from Caucasian to African, and it can be shown that probably 30% of alleles carried by Afro-Americans are of Caucasian origin, which suggests a gene flow of approximately 3% per generation. Population movements by travel, migration, or invasion can alter the gene pool of both the contributing and the receiving populations.

Mutation

A mutation is a change in genetic material, either in chromosomes or in genes (p. 22). Point mutations are those involving changes of the nucleotide bases in individual gene loci that result in a new, different characteristic or set of characteristics, recognized by the sudden appearance of a trait not normal for the species. Mutations may occur in somatic cells and, as such, are not transmitted in the gametes; mutations in the gametes can be passed to future generations. As a rule mutations are not beneficial, although they may in some instances be useful in facilitating adaptation to a changed environment. If the change increases the organism's chances of survival, the new gene is likely to be incorporated into the gene pool, and consequently its frequency will increase. If, on the other hand, the new characteristic produces a change that is incompatible with life (lethal gene), reduces fertility, or so interferes with the fitness of the organism as to be incompatible with the species behavior or reproductive patterns (abnormal genes), the likelihood of transmitting the altered gene is considerably diminished.

Gene mutation is a rare chemical accident that usually involves one gene at a time. Mutations that occur naturally are referred to as spontaneous. In a large majority of cases gene mutation produces a harmful effect; that is, the mutant gene is responsible for the synthesis of a protein that is different from the original protein. A change in the hereditary material causes an alteration in the normal metabolic processes dependent upon it. In most cases when an individual receives a mutant gene from a parent, he receives a corresponding normal allele from the other parent. If the gene does not impair the reproductive fitness of the individual he will in turn transmit the mutation to his offspring. It is, then, mutation combined with other factors that forms the basis of evolution and helps modify the genetic constitution of the species.

It is estimated that each individual in the population carries at least three to eight lethal equivalents; that is, a gene that would in the homozygous state result in death before reproduction. This is sometimes referred to as the *genetic load* of mutations. In other words, everyone possesses genes that reduce health and fertility in some way to a greater or lesser degree. Mutations may be new in that they occurred in the ovum or sperm that produced the individual or they may be old mutations that were inherited from more or less distant ancestors. The measurement of mutation rates is difficult, especially in relation to recessive genes. The spontaneous mutation rate varies for different loci but is estimated to be somewhere between 1 in 25,000 and 1 in 1,000,000 per locus per gamete, with an average of 1 in 100,000. Estimates for some of the more common disorders are listed in Table 8-2.

Induced mutations. Mutations are usually spontaneous, but they are known to be caused by a variety of external agents. Any agent capable of producing mutation is a *mutagen*. Those that have been shown to induce mutations are temperature, certain chemicals, and ionizing radiation.

Table 8-2. Some estimates of mutation rates*

Condition	Mutations per million loci per generation
Dominant conditions	
Facioscapulohumeral dystrophy	5
Marfan's syndrome	6
Myotonic dystrophy	16
Myotonia congenita	4
Huntington's chorea	5
Achondroplasia	53
Retinoblastoma	23
Neurofibromatosis	15
Multiple polyposis	13
X-linked conditions	
Hemophilia	20
Duchenne muscular dystrophy	55
Recessive conditions	
Limb-girdle muscular dystrophy	60
Deaf-mutism	45
Phenylketonuria	25
Albinism	28

*From Porter, I. H.: Heredity and disease, New York, © 1968, The Blakiston Division, McGraw-Hill Book Co. Used with permission of McGraw-Hill Book Co.

TEMPERATURE. Although it has not been demonstrated in man, increased temperature has been shown to increase mutation rates significantly in lower organisms. Of interest to geneticists is the finding that scrotal temperature of clothed males is 3 degrees higher than in nude males, suggesting that this social custom might be a factor in increasing mutation rates in males; there is no evidence to bear this out, however.

CHEMICALS. Numerous chemicals have been found to be carcinogenic (cancer-inducing) and mutagenic in experimental animals. Most of these chemicals (including formaldehyde, phenol, nitrous acid, nitrogen mustard, and other alkylating agents) do not come into contact with germ cells; therefore, the mutagenic effects on man are still only speculative. However, any chemical known to be carcinogenic might also be mutagenic and should be used with care during the reproductive period.

IONIZING RADIATION. The most potent mutagenic agent known is ionizing radiation. The number of mutations produced by radiation is proportional to the total amount of radiation received, whether the exposure is all at once or distributed over a period of time. All body cells can be affected by radiation, but from a genetic standpoint only the dosage to the gonads is of any importance. Direct gonadal irradiation produces greater exposure than the scatter effects from irradiation to more distant body parts; therefore, narrow beams and adequate lead shielding are employed to decrease gonadal exposure during irradiation. Sufficient exposure to high doses of radiation will cause sterility, but the amount necessary to destroy the sex cells would be lethal if applied to the whole body.

The principle sources of radiation to which the human population is exposed are (1) *natural (background) radiation* from cosmic rays, radioactive materials, and naturally occurring radioisotopes that may be ingested or inhaled; (2) *diagnostic* and *therapeutic x-rays,* which are the largest man-made sources of radiation; and (3) the total of *fallout from nuclear explosions* and radiation from *atomic wastes,* the total of which is relatively low compared with background radiation. The units of measuring radiation dosage are the roentgen (r), the unit of exposure dose, and the rad (R), the unit of absorbed dose. It is estimated that the accumulated gonadal dosage received by persons in the United States from conception until the end of the reproductive period averages 3R to 4R from background radiation, approximately the same amount (3R) from diagnostic and therapeutic radiation, and about .3R from fallout. Any amount of radiation, no matter how small, *may* cause a mutation in the genetic material. Awareness of the genetic risks from radiation, man-made in particular, has led to careful use of x-rays and the elimination of many diagnostic procedures when possible.

Selection

Darwin used the term "survival of the fittest" to describe the ability of an organism to adapt to its environment and reproduce its own kind. When a mutation proves to be advantageous the mutant gene will become the prevalent type. This is one of the most important factors in evolution. The forces that operate to increase or decrease fitness take place through the action of *selection* on new genotypes that arise by mutation or recombination. Fitness in a genetic sense describes the ability to transmit genes to the next generation, and is a measure of both survival and fertility. Selection can operate at any time from conception through the reproductive period. Both early death and complete sterility have a common genetic survival. If a dominant mutation is lethal it will be eliminated from the gene pool immediately because its possessor is unable to survive; if it causes sterility the gene will also be eliminated because the individual possessing it will be unable to produce offspring. A recessive gene will be transmitted by normal heterozygotes through generations of offspring until the possessor mates with a similar heterozygote. The gene in double dose will be lethal to the offspring who inherits it, and only in this way will the genes be eliminated from the gene pool. Selection does not act on specific genes (the genotype) but on the individual possessing them (the phenotype).

Many mutations are not lethal, but serve to reduce the fitness of an individual by reducing the average number of offspring produced in relation to individuals who do not possess the gene. In this way, over a period of several generations the gene tends to disappear. The fewer children produced, the more rapidly the gene will be eliminated in a population. The more rare a gene is the less likely will be the chance of its combining with a similar gene; but, slowly or rapidly, deleterious genes have a tendency to disappear from the gene pool. For example, persons with achondroplasia, a form of dwarfism due to a dominant gene, seem to enjoy good health but produce only one fifth the number of surviving children as do their normal siblings. Also, it appears that the fertility of persons with neurofibromatosis (p. 161) is only half that of the population as a whole, possibly due to severe central nervous system involvement that often accompanies the disorder. In addition, any disorder that impairs the mental ability of an individual will reduce the prospects of marriage and thus the probability of offspring. When there is competition among individuals, some are eliminated discriminately.

Sometimes a mutant gene will persist in a population because it increases the fitness of the possessor over that of the normal genotype in a particular environment. The best and most classic example of a selective advantage is the gene for hemoglobin S, which has been described previously (pp. 22 and 63). In double dose the gene is harmful, producing the severe sickle cell anemia from which the affected individuals rarely survive in medically substandard situations. In the heterozygous state (the sickle cell trait) the mutant gene increases the survival rate of its carrier in areas where *falciparum malaria* is prevalent. Persons with sickle cell trait are highly resistant to this form of malaria. In this environment the heterozygote is at an advantage, while the normal homozygote is at risk from malaria, and the abnormal homozygote will die from anemia. There is a high frequency of the gene in the areas where malaria is prevalent; however, in an environment where malaria is not a threat, the deleterious gene will tend to decrease because its usefulness is obviated and only the negative aspects remain.

When a mutant gene is advantageous only up to a point in the special environment, such as the gene for HgS, the gene and its normal allele will attain a state of equilibrium. The balance exists because of the superior fitness of the heterozygote over either the normal or the abnormal homozygote. A major gene and its allelomorph when selectively balanced create a *balanced polymorphism*.

In other recessive disorders, such as cystic fibrosis with an estimated carrier rate of 1 in 20 to 25, it is felt that there may be some as yet undetermined selective advantage conferred on the heterozygote, a condition called *heterosis,* and something else in the genetic make-up that accounts for the persistence of the harmful genes in the population. For example, parents of children with cystic fibrosis seem to have more children; thereby increasing the chance that the gene will survive. Some factors are felt to have influenced the prevalence of diabetes mellitus in the population before the advent of insulin therapy: for instance, diabetic women have an earlier menarche, enabling them to procreate at an earlier age; and one investigator submits that the tendency toward obesity in prediabetics was a decided advantage during times of famine and so contributed to survival of the disease.[8]

Selective forces may be either *natural* or *artificial*. Natural selection takes place under natural conditions and permits the survival and reproduction of phenotypes with certain features that prove useful in their particular environment. Artificial selection is similar to natural selection except that the selected variants are a direct consequence of man's intervention. Effective treatment for otherwise lethal conditions tends to increase the frequency of some genes in the population. For example, the negative selection on such conditions as pyloric stenosis, diabetes mellitus, and retinoblastoma has been relaxed through modern surgery and medical management. Anything that alters the selection against a detrimental gene will tend to increase the frequency of that gene in a population.

Mutation-selection equilibrium

Both selection and mutation operate to alter gene frequencies in a population. When detrimental genes are constantly being eliminated from the gene pool by selection and new ones are being added by the process of mutation, the population is thought of as existing in a state of equilibrium. For example, natural selection is removing gene *a* from the gene pool; but whenever gene *A* mutates to its allele *a,* the mutation process has balanced the selection process. Thus, gene *a* is maintained at a constant level in the population. The more a condition interferes with the affected individual's ability to reproduce, the larger is the proportion of cases due to fresh mutations.

The process of equilibrium can be illustrated by a bird-feeder analogy.[7] If fresh seed is added at the top of the feeder at the same rate the birds are eating seed from the bottom, a constant level of seed will be maintained in the feeder. Assume that the seed in the feeder represents the genetic load, the seed being added is the production of new mutations, and the birds represent the forces of selection. The level of seed in the feeder can be increased in two ways: (1) seed can be added faster than it is being eaten by the birds, analogous to increasing the rate of mutations (for example, increasing the amount of radiation); (2) the number of birds eating the seed can be reduced, analogous to reducing the forces of selection (for example, dietary treatment of phenylketonuria, insulin therapy in diabetes, or surgical correction of pyloric stenosis). As a result of either of these mechanisms, the proportion of deleterious genes is increased in the gene pool.

EVOLUTION

Organisms evolve by adapting to their environment. The diversity of forms seen in the world today was not always as it is today but gradually evolved from very different-looking ancestors. This process is still under way. Children in this generation are able to withstand more environmental pressures than could their parents. The genetic constitution of each new generation differs from that of the one preceding it. Darwin called this process "descent by modification." These modifications result from successful reproduction of individuals with different and distinct heritable characteristics.

All of the factors that operate to alter gene frequencies in populations direct the course of evolution, particularly mutation and natural selection. Of these two, natural selection is the foremost instrument responsible for significant evolutionary changes. Individual differences exist in every species and population, and by the selective elimination of some of these traits the adaptability of the species is assured. Evolution is a gradual change in the hereditary composition of the species. Whereas heredity is a conservative process leading to similarities in familial descent, evolution produces change. Mutations that improve the survival value of the species will be passed to descendants. Even the harmful effects of dominant genes are reduced by modifying genes. Evolution is essentially "the origin of variation and the modification of the variation by natural selection."[6]

Features of evolutionary effects are possible only by comparative studies of

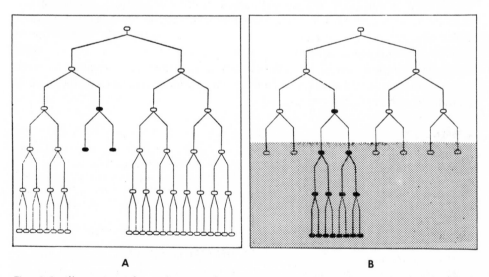

A   **B**

Fig. 8-3. Illustration of a selective advantage. **A,** In a normal environment the black mutants are rare. **B,** In a changed environment the black mutants are better able to adapt than the white; thus they become the predominant types.

past events as practiced by paleontologists and comparative anatomists or by the reconstruction of the evolutionary process through experiments with mechanisms operating in the present as used by geneticists and ecologists. An illustration of the selective process and the production of evolutionary changes in the colon bacteria *Escherichia coli* is demonstrated by a classic laboratory experiment.[1] This organism is easy to culture and divides every 20 minutes to yield billions of descendants in a day. When several billion of these bacteria are placed in a medium containing the antibiotic streptomycin, only a few cells will survive. However, these hardy survivors, streptomycin-resistant mutations, are able to multiply freely in the changed environment (Fig. 8-3). In this way organisms become resistant to current antibiotic therapy so that medical science is continually faced with the task of developing new antibiotic agents to cope with an ever-increasing variety of mutant strains.

A similar example of the effects of altered resistance to an organism is furnished by rubeola (measles). This highly contagious viral disease is seldom fatal and for some time has been considered a universal disease of childhood. However, it is postulated that it was not always so benign (the character of some serious sequelae places doubts on the present status), that at one time this disease claimed as many lives as the more severe disease smallpox, but over a period of 10 generations it became a totally different disease in the populations continually exposed to the virus. When other populations were exposed to the virus they experienced a violent response. For example, the King and Queen of Hawaii died of measles during a visit to England in the nineteenth century, and the virus carried to isolated communities by missionaries created high mortality in these susceptible populations.[1]

Different populations in a species, when subjected to different environmental conditions, will adapt to these varied environments, and in the process of adapting will become genetically different. The diverse changes manifested in the accumulation of distinguishable phenotypic variations constitute the formation of races. When differences persist to the point that these variations become more divergent and so different genetically that members can no longer interbreed, they can be considered to be separate species.

REFERENCES

1. Dobzhansky, T.: The genetic basis of evolution, Sci. Am. **182**:32, January 1950.
2. Edwards, J. H.: Should diabetics marry? Lancet **1**:1045, 1969.
3. Glass, H. B.: The genetics of the Dunkers, Sci. Am. **189**:76, August 1953.
4. Knudsen, A. G.: Genetics and disease, New York, 1965, The Blakiston Division, McGraw-Hill Book Co.
5. McKusick, V. A.: Human genetics, ed. 3, Englewood Cliffs, N. J., 1969, Prentice-Hall, Inc.
6. Mettler, M., and Gregg, G.: Population genetics and evolution, Englewood Cliffs, N. J., 1968, Prentice-Hall, Inc.
7. Moody, P. A.: Genetics of man, New York, 1967, W. W. Norton & Co., Inc.
8. Neel, J. V.: Diabetes mellitus; a thrifty gene rendered detrimental by progress? Am. J. Hum. Genet. **14**:355, 1962.

GENERAL REFERENCES

Crispins, C. G., Jr.: Essentials of medical genetics, New York, 1971, Harper & Row, Publishers.
Emery, A. E. H.: Heredity, disease, and man,

Berkeley, 1968, University of California Press.

Hirshhorn, K.: Errors of metabolism in children, Hosp. Med. **5:**77, 1969.

Levine, L.: Biology of the gene, ed. 2, St. Louis, 1973, The C. V. Mosby Co.

Spar, I. L.: Genetic effects of radiation, Med. Clin. North Am. **53:**965, 1969.

Stern, C.: Principles of human genetics, ed. 3, San Francisco, 1972, W. H. Freeman and Co., Publishers.

Thompson, J. S., and Thompson, M. W.: Genetics in medicine, Philadelphia, 1966, W. B. Saunders Co.

9

Management of inherited disorders

It is apparent from the preceding chapters that inherited disorders constitute an important health problem with far-reaching consequences for individuals, families, and society. The rapid advances made by medical science in the control of and subsequent decline in diseases due to nutritional deficiency and infection have placed diseases in which genetic factors predominate in a new perspective. The total infant mortality, primarily due to upper respiratory tract infections and diarrheal disease, has fallen from approximately 100 in 1,000 live births at the turn of the century to the present 25 in 1,000 live births, of which more than half are attributed to congenital malformations. At the same time, contributions from the fields of biochemistry and cytology have clearly established a genetic basis for an increasing number of diseases. Estimates vary but it is felt that close to 5% or 6% of all live-born infants will manifest a heritable disorder at some time in their lives. Consequently, more and more of the problems encountered by health personnel are related to genetically influenced conditions.

New technologies for identification, treatment, and control of hereditary defects are assuming greater significance in the field of science. The understanding of genetic disease, with the increased survival of affected individuals, has generated concern for the alterations in the genetic load of defective genes and their effect on society and the future of populations, with ramifications extending to economic, ethicolegal, and moral issues. Who will assume the financial burden and responsibility for care of these increasing numbers of affected persons? Should persons with heritable defects be prevented from producing offspring? Who should decide what constitutes a defect? This chapter is concerned with some of these areas of prevention and treatment of inherited disorders, with the major emphasis on genetic counseling.

GENETIC COUNSELING

With the increase in the proportion of genetically determined disorders, the need for competent genetic counseling is assuming greater importance in health education and practice. An increasingly well-informed public is creating a justified demand for accurate information regarding risks to present and future generations. There are at present too few qualified persons to provide families with the information they need regarding their particular problems and the guidelines for a plan of action. Consequently, a growing number of professionals, willingly or unwillingly, assume the counseling role. When expert counseling is not provided, individuals may become the victims of well-meaning but uninformed quasi-professionals and acquaintances. The actual number of persons needing advice is relatively small when compared with those who have many other health problems, but their need is great. Health professionals who are familiar with the facilities in their areas where genetic counseling is available are able to direct individuals and families to needed services.

Genetic counseling services are provided by the major universities and medical centers, and most of them have clinics for the diagnosis and treatment of genetic diseases and birth defects. Many such services are also associated with the large children's hospitals. There are several sources where an up-to-date list of genetic counseling services can be obtained. The National Genetics Foundation* maintains a Network of Genetic Counseling and Treatment Centers throughout the United States and Canada to provide patients and their families with counseling services. The National Foundation–March of Dimes† (abbreviated The National Foundation) not only publishes a Directory of Genetic Services but contributes significantly to professional education and research related to genetic defects and provides extensive programs directed toward teaching and disseminating genetic information. A directory of genetic counseling centers is issued by some state public health departments, and information is available through the Bureau of Maternal and Child Health of the Department of Health, Education, and Welfare in Washington, D. C. Other sources include the numerous organizations concerned with birth defects in general; for example, The National Society for Crippled Children and Adults, Inc. and organizations concerned with specific genetic diseases, such as The National Cystic Fibrosis Society and The Muscular Dystrophy Association of America.

The genetic counselor

There is disagreement among experts as to precisely who should give genetic advice. There are those who feel very strongly that an M.D. or Ph.D. degree is the minimum requirement for a genetic counselor. Others are less adamant about these advanced degrees, but there is consensus that anyone assuming such a role must be highly trained in the principles of genetic theory and the various modes of inheritance, plus have an extensive familiarity with the literature related to heritable

---

*National Genetics Foundation, Inc., 250 West 57th Street, New York, N. Y. 10019.
†National Foundation–March of Dimes, 800 Second Avenue, New York, N. Y. 10017.

factors involved in countless diseases and disorders. The specialist who diagnoses a rare disorder seldom has the additional training to fulfill the role of counselor. He recognizes the need but is unwilling or unable to devote the time and effort necessary to study this additional specialty, nor does he have the time to spend with clients in lengthy counseling sessions. The family doctor who has established a position of trust with the patients under his care is probably in the best position to provide needed counseling. He knows the family and their special needs and their ability to comprehend the information, and he usually maintains long-term contact with the family.

A very few genetic specialists are recognizing the value of resources to be found among other members of the health team. The social worker and the professional nurse, who already have skills in counseling and guidance, might elect a clinical specialization in genetic theory to help fill the need for genetic counselors. They not only count counseling among the tools of their trade but are also in a position to maintain long-term contact with families.

There are increasing opportunities for professionals in other areas of health care to become members of a genetic team. All persons involved in the care and habilitation of affected individuals and their families (nurses, physical therapists, occupational therapists, health educators, teachers of handicapped children, guidance counselors, and so on) would benefit by knowledge and understanding of the mechanisms of heredity and hereditary disorders. Public health nurses are in a unique position to work with families in a close, sustained relationship. Their contact provides the opportunity to reinforce the family's comprehension of information given by the diagnostician and geneticist about diet, medication, and so forth, to maintain followup evaluation of therapies, and to be alert to clues that indicate genetic disorders in unsuspected cases.

Who seeks counseling?

A young couple contemplating marriage is concerned about a suspected genetic disorder in one of their families or they may need advice about cousin marriage. Couples planning adoption may seek advice regarding a prospective child. Some couples may need information about sterilization, artificial insemination, or termination of a current pregnancy. Delayed or abnormal sexual development or recurrent abortions bring a family to the attention of the genetic counselor. He is frequently consulted in cases of disputed paternity. Situations of infertility in one of the marriage partners sometimes involve the genetic counselor. Most infertility of a genetic etiology is related to chromosome abnormalities (for example, XXY, Klinefelter's syndrome); in these cases the best the counselor can hope to do is explain the disorder and offer psychologic support. Occasionally the geneticist is called on to establish zygosity of twins, usually in relation to possible tissue donors.

Most often persons seeking genetic advice are parents of a defective child who are concerned about having another similarly affected child. There is rarely a simple answer to their questions. Each case must be evaluated on an individual basis to determine the risk related to the specific situation. The information is presented to the family in a way that is easiest for them to comprehend; then the decision is left up to them.

Objectives of genetic counseling

Carter has outlined three essential objectives of genetic counseling: (1) to advise parents and answer questions in regard to risks of recurrence of a genetically determined abnormality, (2) to alert the medical profession to the possibility of a genetic disorder in an unborn child, and (3) to reduce the number of children who are born with a genetic disorder.[1]

Advising parents. More than ever before, parents plan and feel responsible for their children. They are becoming increasingly knowledgeable about genetics and the adverse effects of environment on the unborn child, with an attendant demand for information. Parents are entitled to accurate information regarding risks, to have this information presented in language they can understand and in the proper perspective. That is, they need to know the risks related to their particular situation compared with the random risk for any prospective parents. For example, a risk of 1 chance in 20 is less threatening in relation to a 1 in 30 chance for any random pregnancy. The prospects of an early, inevitable death is very different from life-long disability or a disability for which there is a remedy. It has been found that when parents understand the risks involved they normally make sensible decisions regarding family planning.

Special risk situations. When the attending physician is alerted to the possibility of a heritable disease in a family, this knowledge facilitates the early detection and subsequent treatment of the disease. This is increasingly important as more treatments are becoming available for genetically determined disease and is especially true in situations where treatment is effective only when initiated early. History of a condition in an older sibling, such as phenylketonuria, galactosemia, or nephrogenic diabetes insipidus, provides a clue to the pediatrician for specific and thorough testing for the condition in a newborn. In this way an early dietary regimen can be initiated when indicated, thus minimizing or eliminating any deleterious effects of the disease. As more biochemical defects are recognized, techniques for multiple testing will facilitate early detection of numerous metabolic diseases in the newborn period.

In situations of questionable or tenuous diagnosis, family history of a hereditary disorder often provides direction for the attending physician. For instance, history of Hirschsprung's disease in a family may alter directions for the diagnosis of constipation; history of cystic fibrosis will be an indication for intensive testing in cases of chronic upper respiratory tract infection in early infancy; history of gout may provide clues to a perplexing case of arthritis.

Reducing numbers of affected children. Now that the heterozygote can be identified in an increasing variety of single-gene defects, this aspect of genetic counseling is assuming greater importance and offers hope in preventing disabling disease. Persons with a family history of some hereditary disorders are able to ascertain before initiating a pregnancy whether they are carriers of the gene for a severe defect. In these instances the genetic counselor is able to advise the couple on the risk related to any pregnancy, in regard to the specific defect, with a high degree of accuracy. New techniques for prenatal diagnosis of chromosomal aberrations and an increasing number of metabolic defects have created a means for detecting the presence of an abnormal fetus early in pregnancy, providing a couple with the prospect of a defective child the option to terminate the pregnancy.

FUNDAMENTALS OF GENETIC COUNSELING

In a medical prognosis the physician attempts to make a prediction regarding the outcome of an illness. A genetic prognosis is not only a prediction of a disease outcome but may apply to relatives and future offspring of the person seeking advice. The person who seeks genetic advice (the *consultand*) may not necessarily be ill himself, but concerned about the effects a hereditary disorder might have on future children. Effective genetic counseling is based on information derived from several sources: accurate diagnosis, extensive family history, and a sound knowledge of genetic principles and background of the literature.

Accurate diagnosis

An accurate diagnosis is absolutely essential in order to provide information regarding probability of recurrence. This is particularly important in relation to genocopies or mimic genes. For example, sometimes the various forms of muscular dystrophy have similar manifestations but the modes of inheritance are totally different (p. 66). The facioscapulohumeral form is inherited as autosomal-dominant, which means that the affected person can transmit the disorder to 50% of his children. The limb-girdle variety is autosomal-recessive and thus manifest only in the homozygote; there is risk to the children only if his mate carries the same defective gene. The X-linked Duchenne type, transmitted by the female line, is a threat to male children. Similar disorders with different modes of inheritance are the mucopolysaccharidoses: the X-linked Hunter's syndrome and the autosomal-recessive Hurler's syndrome.[1, 10]

Not all disorders affecting more than a single member of a family are hereditary. A good illustration of such a situation is pellagra, a disease due to a dietary deficiency of niacin, characterized by a typical dermatitis and frequently by nervous system involvement. Its prevalence in several members of the same family and often in two or three generations gives the false impression of a hereditary disorder although it is a result of the shared diet. Other obvious environmental causes are food poisoning and many infectious diseases. When a disorder is found in definite proportions in families (for example, persons related by descent) and absent in unrelated individuals (spouses, in-laws) the likelihood of a genetic disorder must be considered.

There is a characteristic age of onset associated with many heritable disorders that provides a valuable clue to etiology in the absence of other precipitating factors (Table 9-1).

Family history

A careful and detailed history of the family is the basis for any form of genetic counseling. Not only does it give a picture of the disorder in relationship to other members of the kindred, but by observation of family patterns it may serve to confirm a tentative diagnosis. The history can identify the genetic makeup of the par-

Table 9-1. Characteristic onset of some genetic diseases*

Age of onset	Condition
Lethal during prenatal life	Some chromosome aberrations Some gross malformations
Present at birth	Congenital malformations Chromosomal aberrations Some forms of adrenogenital syndrome Some forms of deafness
Soon after birth	Phenylketonuria Galactosemia Sometimes cystic fibrosis
Infancy	Tay-Sachs disease Werdnig-Hoffman disease Maple syrup urine disease
Early childhood	Cystic fibrosis Duchenne muscular dystrophy
Near puberty	Limb-girdle muscular dystrophy Some forms of adrenogenital syndrome
Young adulthood	Acute intermittent porphyria Hereditary juvenile glaucoma
Variable onset age	Diabetes mellitus (0 to 80 years) Facioscapulohumeral muscular dystrophy (2 to 45 years) Huntington's chorea (15 to 65 years)

*Data from Porter, I. H.: Heredity and disease, 1968, The Blakiston Division, McGraw-Hill Book Co.; and Thompson, J. S., and Thompson, M. W.: Genetics in medicine, Philadelphia, 1966, W. B. Saunders Co.

ents (or parent), and in more complex multifactorial inheritance will help to determine the level of risk for a particular patient. Detailed information is needed about the proband's relatives even when he is obviously suffering from a defect that is clearly genetic with an established mode of inheritance. For example, a patient with a confirmed diagnosis of osteogenesis imperfecta exhibits gross bone deformity. Superficially the family appears to be unaffected, but a careful history will elicit the presence of blue sclera in the father and a paternal aunt with otosclerosis. This illustrates the deceptive nature of variable expressivity in a dominant disorder and the way that symptoms are often overlooked unless careful questioning is carried out.

The person taking a family history must allow a liberal amount of time. In the case of an affected child, the interview should include both parents in order to elicit information about relatives on both sides of the family. Medical records and family Bibles are helpful sources. Other relatives may be consulted, and grandmothers are a goldmine of information. Reliability of the information regarding whether relatives were affected will vary greatly. The manner in which the information is elicited determines to a large degree the extent of information that is forthcoming and the attitude of the informants. There may even be reticence on the part of the informant, particularly if the disorder is considered a "skeleton in the family closet." Sometimes true relationships are concealed, such as illegitimacy. Information concerning first-degree relatives is most important, and the data on these should be complete. It is not uncommon to discover more serious conditions in the family than the one under investigation. For example, while questioning parents about a child with congenital dislocated hip, an investigator found hemophilia B in a sibling of which the attending physician was unaware.

The family history is recorded in diagrammatic form by construction of a pedigree using standard symbols (p. 32). The extent of the pedigree will depend on the disorder but should include the proband, his sibship, and the preceding and succeeding generations. It should contain significant information about each member and include live births, stillbirths, and abortions (miscarriages). Situations that concern consanguineous marriages, multiple marriages, or any other complex relationships require extra care in order to outline the relationship accurately. There should be notations about the age (recording the year of birth is best), the date and place of birth, death and cause of death. This facilitates tracing of hospital records and death certificates if needed. The place of birth can often be significant. The incidence of Tay-Sachs disease in Ashkenazic Jews is significantly higher than in those from other areas; thalassemia is higher in persons from the Mediterranean region.

Actual construction of the pedigree begins with the proband and the outcome of all the pregnancies of the mother. Any abnormalities of the pregnancies are noted, such as high blood pressure, bleeding, anemia, convulsions, excessive weight gain, x-rays or infectious disease. The outcomes of previous marriages may also be important, and such facts as history of radiation exposure of the father should not be overlooked. Next the medical histories of the maternal relatives are explored, beginning with the mother's siblings and including the outcomes of her mother's pregnancies (stillbirths, miscarriages, and so forth). The fact that two sisters died in infancy as blue babies would be significant, whereas a healthy sibling who died in an automobile accident at age 6 would not. Similarly, a history of an infant who died from "stoppage of the intestines" would be meaningful in the investigation of a child with chronic bronchitis (both are symptoms of cystic fibrosis). Details concerning the general health or death of the maternal grandparents, nieces, nephews, uncles, aunts, and first cousins are included in a family history if the mother has information about them; the same categories of relatives on the father's side are explored. It is important at this point to determine whether the couple might be related in any way.

When a family history is completed, the pedigree will reflect either a positive family history in which other relatives are affected with the same disorder or a negative family history in which the proband is an isolated case. It is at this point that the counselor calls upon an understanding of the principles of genetics, a knowledge of the risks related to multifactorial inheritance, and up-to-date information on genetic diseases in order to counsel families regarding their specific problem.

Application of genetic principles and estimation of recurrence risks

The degree of risk related to the major categories of genetic diseases depends upon the mode of inheritance. The random risk of a defective child for any pregnancy is felt to be in the order of 1 in 30. As a general rule, the more definite and clear-cut the genetics, the greater are the risks; as the genetics become more obscure, the outlook becomes more hopeful. Experts in the field of genetic counseling have determined that recurrence risks for heritable disorders fall into three broad groups: (1) those in which the risk of recurrence is little more than the random risk for any pregnancy in the population, (2) those in which there is a high risk of recurrence of 1 in 10 or greater and more often 1 in 4, and (3) those in which there is a moderate risk of better than 1 in 10 and usually less than 1 in 20.[8]

Random risk situations. Conditions that are due to environmental agents and, therefore, not likely to recur in another pregnancy are considered to be random risks. Examples of conditions in this group are defects resulting from maternal rubella, toxoplasmosis, or ingestion of a teratogenic drug such as thalidomide. In these instances the environmental agent affects the developing embryo, not the germ cells. A subsequent pregnancy would carry no more risk of the same defect than for any person in the general population.

There are a number of instances where the condition is genetically determined but the individual is affected as the result of a fresh mutation. This includes a large number of children with severe dominant conditions and most of the chromosome abnormalities. With a chromosome aberration such as trisomy G (Down's syndrome) the estimate of recurrence would probably be considered to be about double the random risk for another mother of the same age, provided the parents are chromosomally normal (p. 81).

High risk situations. The majority of instances where there is a high risk of recurrence are due to mutant genes that produce a large effect; a minority are due to chromosomal aberrations in which a parent has a similar anomaly. When the condition is due to a factor that segregates, the probability of an affected offspring can be predicted with a high degree of accuracy.

CHROMOSOMAL ABERRATIONS. An example of a high risk situation due to a chromosome abnormality is the D/G translocation Down's syndrome (p. 78). In this disorder a phenotypically normal person passes the translocated chromosome to his offspring. Theoretically, the chances for the birth of a defective child when one parent is a balanced translocation carrier is 1 in 3. Since a high percentage of translocation abnormalities are aborted early in pregnancy, most authorities estimate the risk for an affected child to be in the order of 1 in 5 when the mother is a carrier and less than 1 in 20 when the father carries the translocation.

SINGLE-GENE DISORDERS. Disorders due to a single gene account for the majority of conditions with a high risk of recurrence. Although in most cases the probability is clear-cut, in many there are complex factors that complicate predictions. Probability estimates in *autosomal-dominant* conditions can be predicted whether the consulting parents have had children or not. For a parent who displays a dominant defect such as Marfan's syndrome (p. 44), polydactyly, or achondroplasia, there is a 50% chance that the offspring will have a similar defect. Sporadic cases due to a fresh mutation hold no threat to subsequent children, but the affected child will be able to transmit the gene to half his offspring. Difficulties arise when the disorder is only partially penetrant, shows variable expressivity, or has a characteristic late onset. For example, a 50-year-old man whose father died of Huntington's chorea asks what the chances are that his son will develop the disease. This disabling and ultimately fatal disease of the central nervous system manifests itself between 30 and 50 years of age, with an average of 35 years. This risk to the son depends upon whether the father has inherited the gene from *his* father. There would be a 50% chance that the father had the defective gene and a 50% chance that the son received the gene from the father—an estimated risk of 25%. However, since the father is within the outer limtis of the average age of onset for this disease, the likelihood that he does not have the defective gene is considerably improved. This is taken into account by the genetic counselor. By the use of complex computations based on the probability that 80% of cases are manifest before age 50, the counselor can often arrive at a risk estimate considerably less than 25%.

A similar probability would be operative when the disorder is only partially penetrant; for example, retinoblastoma, which is 80% penetrant. The likelihood would be entirely dependent upon whether the parent has inherited the gene but does not display the defect. If he indeed has the gene in nonpenetrant form there is still a 50% probability that he will transmit the defective gene to his offspring. Nonpenetrance accounts for the skipping of generations. Also, when apparently normal parents have two or more children affected with the same autosomal-dominant condition, the likelihood that both are a result of fresh mutations is remote. A more likely explanation is that the mutant gene is nonpenetrant in one parent. Sometimes very careful examination of the parents will reveal minor evidence of the disorder.

The variability in expression of some autosomal-dominant disorders may also create counseling difficulties. A member of a family with a history of a disorder, such as osteogenesis imperfecta or Marfan's syndrome that may have one of several manifestations, might appear to be free of the disorder but on careful examination prove to carry the gene and exhibit a very minor manifestation. One half of his children would be at risk to display any of the symptoms of the disorder.

Recurrence risk estimates for *autosomal-recessive* disorders such as cystic fibrosis or phenylketonuria are most important in regard to subsequent sibs of an affected child. The affected individual is homozygous and has received a gene from each parent; therefore, the recurrence risk is 1 in 4 of an affected child for each pregnancy. There is relatively little risk to other relatives although there is a two-thirds chance that other sibs of the carrier couple will be heterozygous for the defective gene. Even in cystic fibrosis, with the highest estimated carrier frequency of 1 in 25, the risk that two heterozygotes will marry is relatively slight. Offspring of the affected homozygote will inherit the gene, but again the risk of mating with another individual with the gene is only moderate. Males with cystic fibrosis are sterile, and the risk for affected females producing an affected child is estimated to be approximately 1 in 45.[1]

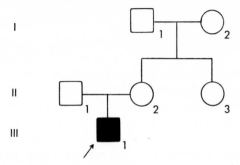

Fig. 9-1. Pedigree of an affected male. See text for details.

Identification of heterozygotes for autosomal-recessive disorders makes risk estimates more precise, and prenatal identification of the disorder offers the option of terminating the pregnancy when a fetus is affected.

For *X-linked recessive* disorders the significant risk is related to male offspring of a female carrier, and is 1 in 2. When an X-linked disorder appears in a family the risks to other relatives are dependent upon whether the females are heterozygotes. For example, a woman whose sister has a son with Duchenne muscular dystrophy (p. 67) is contemplating marriage. She seeks counseling to determine the probability that she might also produce an affected son. The probability that her sons are at risk will depend on whether or not her mother was a carrier of the defective gene (Fig. 9-1). The affected child (III_1) could be the result of a fresh mutation. Because a mutation is usually an isolated case and affects only a single germ cell, there would be no more risk to subsequent sibs than before the first affected son. If the mother (II_2) received the defective gene from *her* mother (I_2) the risk would be 1 in 2 for a second affected son and the same risk that the daughters are carriers. Sons of II_3 would be at risk if her mother (I_2) was a carrier and transmitted the gene to the consultand (II_3), in which case the risk would be 1 in 2 for an affected son. If both sisters had an affected brother the carrier status of I_2 would be assumed.

Moderate risk situations. The largest group of common conditions belong to the moderate risk category. These include the multifactorial disorders with risks less than 1 in 10 but substantially greater than the random risk in the population as a whole. Risk recurrence in these disorders is *empirical.* Empiric risk estimates are not based on genetic theory but on prior experience and observation. They are determined by a knowledge of the frequencies that have been observed in other families with a similar condition as related to the incidence of the disorder in the family under consideration. More and more of this information is available in the literature. Since these risks are only estimates based on incidence in other families, the figures are undoubtedly exaggerated in some families and underestimated in others.[4] The involvement of more than one member of a family usually indicates more multifactorial traits in common. When two siblings are affected the risk to subsequent offspring is doubled. Most empiric risk estimates for common conditions were included in the discussion in Chapter 7 of some of these disorders.

Interpretation of risks

When giving risk estimates the genetic counselor does not attempt to make recommendations for clients, who are usually parents who have suffered the birth of a defective child. The counselor provides appropriate and accurate information about the nature of the disorder, the extent of the risk involved, and the probable consequences, but he leaves the final decision to the parents.

In most instances counseling is a pleasant task and serves to reduce rather than increase anxiety in the parents. For example, a risk estimate of 1 in 25 that a second child might be born with a cleft lip and palate looks much less formidable when it is considered that there is a 1 in 30 risk of some defect for any couple. Likewise, stating a 95% chance that another child will not be affected is more positive than a 5% chance that he will.

Most counselors find that clients understand and readily accept answers and explanations that are given in terms of odds. Few people have not had experiences with flipping coins, baseball pools, lotteries, and weather reports. The odds of heads or tails in a flip of a coin is analogous to the 50-50 probability in a dominant disorder. If heads represents a dominant gene and tails its recessive allele, with the flip of two coins there would be a 1 in 4 chance that the coins would both be tails—analogous to a recessive homozygote. A weather report is a well-known example of an empiric risk estimate.

It is important to impress upon clients that each pregnancy is an independent event. Parents of one affected child cannot assume that a 1 in 4 chance of another affected child assures them that the next three will be normal. "Chance has no memory." The risk is 1 in 4 for each and every pregnancy (just as the probability that the flipping of two coins will produce tails).

A source of added stress occurs when a disease is not diagnosed until the mother is pregnant with another child. If the disorder is one that can be detected by amniotic fluid examination (for example, Down's syndrome or Tay-Sachs disease) the couple is offered this opportunity to remove doubt regarding the unborn child. In this situation the counselor again uses a positive approach. Telling the parents of a child with an inborn error of metabolism such as Tay-Sachs disease that there is a 75% chance that the child does not have the disease is more reassuring than that there is a 25% chance that it has the same defect. Where the results are negative, the couple can continue the pregnancy knowing it holds no more risk than any random pregnancy. In the event that the examination reveals a similarly affected fetus the parents then have the option to terminate the pregnancy and perhaps attempt another.

Consanguinity. The problem of a proposed marriage between blood relatives may bring a couple to the genetic counselor. There is considerable disagreement among authorities regarding whether there is a significant risk of a defective offspring in such matings. The question is especially difficult regarding first-cousin marriage. There is a higher incidence of postnatal mortality and an increased frequency of congenital disorders and mental retardation reflecting common genetic factors. However, the actual risk is relatively small and many couples accept it. Certainly the more genes two persons have in common the greater is the probability that there will be an increase in identical polygenic complexes in their children, thus raising the multifactorial threshold for many common defects.

Counseling against consanguineous marriages is usually on the basis of an increased probability that they will express harmful recessive genes. The chances that a person who carries a rare recessive gene will marry someone who has a similar gene is much greater if he marries a close relative. The coefficient of inbreeding, or the probability that the persons will receive a specific gene from common ancestors (Table 7-1, p. 140), is used by the counselor as a tool for estimating the added risk related to a consanguineous mating. For example, in a common disease such as cystic fibrosis the chance of a carrier mating with another carrier in the general population is 1 in 25. However, the chance that a cousin carries the same gene is 1 in 8, or three times the risk in a random mating. The number of genes in common fall rapidly (by one half for each generation removed from the common ancestor) as the relationship becomes more distant. For example, second cousins have 1/32 of their genes in common and probably carry only a slight risk. Some authorities do not discourage consangineous marriage on the basis that a defective offspring of the union will serve to remove two deleterious genes from the gene pool.

Adoption. Problems involving child adoption are often brought to the genetic counselor. Adoption agencies wishing to place a child may hesitate to do so when there is a suspected hereditary disorder in the child's background. The major difficulty arises in situations where the manifestation of the disease does not become evident until later years, as in diabetes or adult-type polycystic kidney, and there is no biochemical test to identify its existence at an early age. In questions regarding probability of an incestuous union the adoptive child is observed carefully for signs of any heritable disorders.

Followup

Followup of genetic counseling is all too frequently neglected. The success of counseling is measured by the way in which the families utilize the information presented to them. In many disorders a diagnosis of one family member places relatives at risk. For example, all family members should be screened in cases of polyposis of the colon. In a disorder such as phenylketonuria that requires conscientious diet management to prevent disabling complications, it is important to make certain that the family understands and follows the advice. Subsequent children must be carefully checked to detect early symptoms. Families can be directed to agencies and clinics that specialize in the disorder and that can frequently provide services in the form of equipment, medication, and rehabilitation.

Often the family has not really "heard" the information presented to them so that it may be necessary to repeat and reinforce counseling. Maintaining contact with the family or referral to an agency that can provide a sustained relationship is essential to the success of any counseling program.

Psychologic aspects

Counseling is dealing with people—people under stress. It demands time and understanding to deal with the emotional tension and anxiety generated in clients faced with the prospect of a disabling condition. Knowledge of and the ability to deal with the range of psychologic responses with all their ramifications, such as grief reaction, guilt, and coping mechanisms, are essential components of the counseling role (there is excellent resource material available in this area). These factors determine the degree to which a counselor's message is understood and influence the client's attitudes and the use he makes of the information. For example, denial, often a defense against the dashing of hopes, is based on emotional conflict and inhibits understanding in a number of families. Awareness and understanding of these feelings makes the difference between a genetic informant and a genetic counselor.

Self-blame is a very natural and universal reaction. Often the counseling person is in a position to absolve the parents of guilt by explaining the random nature of segregation during both gamete formation and fertilization. There is comfort in knowing that everyone carries defective genes and that it is sheer chance that a particular couple happens to carry the same abnormal gene. Reactions may be different in situations where one member can pinpoint the "blame" (dominant or X-linked disorders), whereas there is some measure of reassurance (in recessive disorders) for the couple to know that it is not just one of them who carries the defective gene.

It is important to stress the fact that there is nothing shameful about an inherited or congenital defect and to emphasize any remedy if appropriate. Families have a tendency to be more ashamed of a hereditary disorder than one caused by self-indulgence such as obesity or alcoholism. The threat of a hereditary taint often creates intrafamily strife, hostility, and marital disharmony sometimes to the point of family disintegration. Relatives frequently cease reproduction after the diagnosis of a hereditary defect. The decision to marry may be deferred on the basis of a disorder (even a remote one) in a partner's family. Parents of one member of the involved couple wish to know about the prospective in-laws. Evidence of an alleged hereditary defect in one family line, founded or unfounded, may serve to induce parental pressure to break up an engagement. Accurate assessment from an expert in the field of genetics can confirm or dispel anxieties for the parties concerned, including old wives' tales and superstitions. In any case, counseling provides the concerned individuals with the information needed to make a decision regarding marriage with or without children.

The manner in which the counselor presents the situation depends a great deal upon the nature of the condition and the burden it may place upon the family. A burden is considered to be the total amount of distress created by the birth of a defective child—the anticipated burden as well as the threat of disability.[3] For example, some parents may be willing to assume the burden of a minor or moderate physical handicap such as a claw hand but be loathe to hazard the burden of a mentally retarded child. The inevitable death from any disorder is distressing, but to lose a child with Tay-Sachs disease at age 3 or muscular dystrophy at 16 is quite different from the loss of an anencephalic child at birth. The prospects of a lifelong disability created by an extra finger or even diabetes would not have the impact of the paraplegia of myelomeningocele or the incapacitating deterioration of spinal cerebellar degeneration or mental retardation. The longer the duration of the illness, the greater the financial and emotional burden.

Persons respond differently to probabilities. A risk that is reassuring to one may be threatening or intolerable to another. On the other hand, two individuals will respond differently to a hazard that both see as threatening. Some parents will choose to have children even in the face of high risk, while others feel that even a moderate risk is too much to take. Parents, who elect to have children in spite of a fairly high risk of recurrence, can be helped by education. By learning about the disorder they will be alert to signs of the disease so that early treatment can be initiated to minimize the ill effects of some disorders.

Some of the obstacles to the use of counseling involve the understanding and attitudes of the clients.[3] A frequent barrier to an objective use of information is religious attitude toward conception and the opposition to sterilization and abortion in situations where prenatal diagnosis has indicated a defective fetus or high risk of recurrence. Another is the question of individual rights. A person with a disabling condition may feel that he is entitled to the same rights as anyone else, including the right to procreate children even in the face of a 50% chance of a similarly disabled offspring.

Differences in the ability to comprehend what is said inhibits comprehension of such concepts as probability. Experience, education, and intellectual level must be considered when interpreting risks to a family. One of the greatest obstacles in relation to understanding information is a lack of knowledge of genetics and human biology. Even though the consultands can repeat information, they may be unable to grasp the genetic characteristic of a specific disorder.

CONTROL OF GENETIC DISEASE

At the present time there is no cure for genetic disease although preventive methods and treatments are helping to reduce the harmful effects in an increasing number of conditions. Active research in the area of gene therapy, the treatment of hereditary disease by influencing the genes directly, may ultimately provide solutions to problems in management of genetic disease, perhaps at the risk of creating new problems.

Treatment of genetic disease[4, 5]

Methods of therapy exist for many genetic disorders, due in part to the fact that most all diseases are the result of the interaction of both genetic and environmental elements. An overview of some of these techniques includes the following.

1. *Diet.* If accumulation of a substrate (a substance upon which an enzyme acts) is responsible for adverse effects then, logically, restriction of the substrate (or its precursor) would prevent undesirable symptoms. For instance, restriction of substrates is utilized in the enzyme deficiency diseases such as phenylketonuria (p. 50) and galactosemia (p. 52), in which elimination of foods containing

phenylalanine and galactose prevent irreversible damage from these improperly metabolized compounds.

2. *Product replacement*. In some deficiency diseases, supplying the missing factor that cannot be synthesized prevents undesirable effects. For example, thyroid extract is provided to prevent cretinism due to a genetic defect in the synthesis of thyroid hormone, and the administration of missing blood factors can reduce life-threatening and disabling hemorrhages in the hemophilias. Other examples include corticosteroids for adrenogenital syndrome and insulin in diabetes mellitus.

3. *Avoidance of drugs*. In drug-induced disease such as glucose-6-phosphate dehydrogenase deficiency and the porphyrias, avoidance of the drugs that precipitate a reaction provide a simple preventive measure.

4. *Removal of toxic substances* that accumulate in vital tissues as a result of a hereditary disease can prevent disabling complications. Some of the deleterious effects of hemochromatosis, a hereditary disorder characterized by an excess accumulation of iron in the liver, heart, and pancreas, can be reduced with the removal of iron from the body by periodic venisection. Excess copper that accumulates in the liver and brain in Wilson's disease can be removed by combining the copper with certain drugs.

5. *Transplantation*. Replacement of nonfunctioning organs with normal ones will increase survival of affected individuals once the problems of tissue incompatibility are overcome. Examples include replacement of hereditary polycystic kidneys with normal organs, and bone marrow transplantation as a possible therapy in hereditary diseases affecting the blood-forming organs, for example, thalassemia.

6. *Surgery*. Surgical repair of defective organs can prolong life in multifactorial disorders such as congenital heart disease and pyloric stenosis. Surgical removal of the colon in cases of polyposis coli eliminates the countless polyps that invariably become cancerous. Splenectomy (removal of the spleen) prevents the trapping of abnormal blood cells in that organ in the hereditary anemia spherocytosis.

7. *Immunologic prevention*. The administration of immunoglobulin to Rh-negative mothers following birth is effective in preventing Rh-antibody formation that causes hemolytic disease of the newborn in subsequent births (p. 122).

8. *Enzyme induction*. In some metabolic diseases that have only a portion of the normal activity of an enzyme, stimulating the body to increase total production will bring the enzyme level near normal. To illustrate, the drug phenobarbital induces the production of enzyme in a form of hereditary jaundice due to a defect in an enzyme necessary for the conversion of bile pigment for excretion.

9. *Cofactor supplementation*. The reaction rates of some enzymes are enhanced by vitamins that serve as coenzymes, or cofactors. In the absence of or the inability to synthesize these cofactors from precursor substances, increased amounts of vitamin must then be an essential component in the diet. Some inborn errors of metabolism prevent the effective utilization of these substances due to lack of some essential component. The multifactorial disorder pernicious anemia is due to lack of mucoprotein in the stomach mucosa, which prevents the absorp-

tion of vitamin B_{12} in sufficient amounts needed for the maturation of red blood cells. The anemia is prevented by the administration of vitamin B_{12} in large amounts orally or parenterally.

10. Other methods such as *enzyme repression* and *competitive inhibition* are providing effective treatment in some metabolic disorders. More of these techniques are likely to be instigated in treatment of genetic disorders as advances in biochemistry determine the basic defects in increasing numbers of diseases.

Prevention of genetic disease

In addition to medical treatment, other methods unique to genetic disorders are employed primarily in prevention of disease. The major method is *genetic counseling* to advise relatives on risks or recurrence of a genetic disorder and the likelihood of occurrence in a marriage of two heterozygotes. Genetic counselors, although nonjudgmentally neutral, do indeed exercise an influence over the client's decision, and most find that couples faced with high risk of defective offspring elect to have fewer children than they otherwise would. Preventive techniques used in conjunction with counseling are heterozygote detection, intrauterine diagnosis, and selective abortion.

Heterozygote detection. Tests for detection of carriers of genetic disease is assuming greater importance as more defects are identified and techniques are developed for easy application. Mass screening for numerous defects may eventually be a routine procedure just as screening is now carried out for some specific diseases such as phenylketonuria. Should this become a reality, in theory autosomal-recessive phenotypes might be eliminated.

Although not a preventive technique, detecting the presence of a mutant gene in a healthy individual provides him with this knowledge for use in family planning. Carriers of many inborn errors of metabolism can be identified by laboratory tests but, because of their rarity, mass screening is unfeasible except in persons known to be at risk. Persons at risk include relatives of affected persons, such as sibs of an individual with an inborn error of metabolism (parents are assumed to be carriers), who have a 2 out of 3 likelihood of being heterozygous, or sisters of a male with an X-linked disorder, with a 50% chance of being a carrier. Others at risk are persons whose ancestry or geographic location place them at risk; for example, the carrier rate of sickle cell anemia is estimated to be 1 in 10 for blacks, and 1 in 30 Ashkenazic Jews are carriers of the gene for Tay-Sachs disease.

To be truly effective, screening programs must be accompanied by counseling and education. Merely being aware of the existence of the gene is not enough; families need to know what this means in terms of their own lives. It is particularly important when screening for genes, not diseases, in persons who are well but who may be at risk to produce a disease for which no prevention, prenatal detection, or treatment can be offered for affected individuals (such as sickle cell anemia).

Intrauterine diagnosis and abortion. Current technology provides the ability to establish a diagnosis of genetic disease in a fetus, thereby predicting the risk of abnormality in a newborn. The procedure (amniocentesis), performed at 14

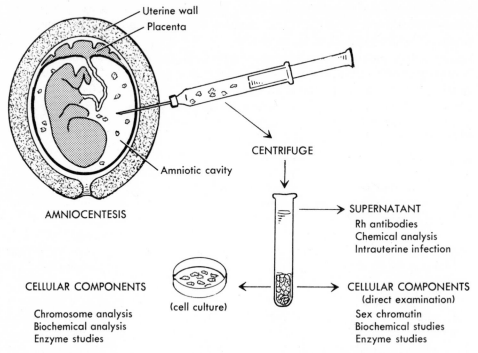

Uterine wall
Placenta

CENTRIFUGE

Amniotic cavity

AMNIOCENTESIS

SUPERNATANT
Rh antibodies
Chemical analysis
Intrauterine infection

CELLULAR COMPONENTS

(cell culture)

CELLULAR COMPONENTS
(direct examination)

Chromosome analysis
Biochemical analysis
Enzyme studies

Sex chromatin
Biochemical studies
Enzyme studies

Fig. 9-2. Amniocentesis and laboratory utilization of amniotic fluid aspirant.

to 16 weeks' gestation with the mother under local anesthesia, consists of removing some of the amniotic fluid that surrounds the fetus during intrauterine life in order to study its contents. The fluid constituents (fetal urine, respiratory secretions, and viable cells shed from fetal skin and respiratory tract) are used for cytologic studies, enzyme studies, or biochemical analysis. The aspirated fluid is centrifuged to separate the cellular components from the fluid, and each can be used for examination (Fig. 9-2).

Indications for amniocentesis are:[6] (1) chromosome disorders in which a parent is a chromosome mosaic, has a balanced translocation, or is a woman 40 years of age or older; and (2) single-gene disorders where both parents are heterozygous for an autosomal-recessive disorder, the mother is a carrier of an X-linked disorder, or the couple has already had an affected child. In cases where the fetus is found to be affected or is a male fetus of an X-linked carrier, the parents are offered the opportunity to abort the affected fetus. In most instances prenatal diagnosis is performed only when the couple agrees to abortion of an abnormal fetus; otherwise, the test would serve no useful purpose. The question of abortion, for any reason, always involves legal and ethical issues that will not be elaborated upon here.

From a geneticist's point of view, the selective elimination of affected fetuses through abortion, thus allowing parents to complete pregnancies with phenotypically normal children, might serve to increase the frequency of some genes in the gene pool.[2] For example, a couple who are known carriers of a disease such as cystic fibrosis and who want two children will determine at each pregnancy whether the child will or will not be affected. The prevention of natural selection against the deleterious genes through death or no reproduction of the affected homozygote will increase the gene frequency in the population over numerous generations by way of increased transmission of the gene by heterozygous gene carriers. At least one authority submits that if all pregnant women over 40 years of age had an amniocentesis, with selective abortion of affected fetuses, about 20% of cases of Down's syndrome would be prevented.

Future management of genetic defects

Future therapy for hereditary diseases will probably include enzyme replacement by pills or injections, the transfer of genetic material into the cell by way of a nonpathogenic virus carrier (transduction), and enzyme stabilization by chemical or physical agents. A number of ways have been envisioned for alteration of germ cells, including directed mutagenesis to reverse a mutation that produced a defective gene. Replacing large specific segments of DNA is proposed as a highly unlikely means to improve polygenic behavioral traits.

In vitro fertilization of human eggs, obtained by a simple laparoscope technique, with sperm from a donor (either the husband or another male) will be possible. These fertilized eggs, analyzed for genetic defect, can be reimplanted in the egg donor or any female for development and delivery. Predetermination of sex will probably be a reality in the near future. Cloning, or the process of removing the nucleus from a germ cell and replacing it with a diploid nucleus from a somatic cell to produce an exact copy of the donor, has been successfully accomplished in frogs.

All of these and other projected feats of genetic engineering will undoubtedly create as many or more problems than they solve. It is predicted that science will provide the means to alter genes before society is ready to respond appropriately.

GENETICS AND SOCIETY

Almost every society has established regulations designed to improve the genetic composition of the society. Laws that prohibit marriage between first-degree relatives is universal, and some societies prohibit marriages between members of different racial groups. Most have laws against marriages of mentally deficient persons. In some primitive societies the chief who gains his position by virtue of superior endowment is permitted the largest number of wives. In many there has been established a hierarchy of expendable individuals for selective elimination during times of famine or limited resources.

Social forces have influenced genetic structure, and the genetic constitution of a society has influenced its social structure and economy. For example, the hereditary deficiency of an intestinal enzyme that is necessary for the digestion of milk sugar may be responsible for the lack of dairying as a food source in some Asian and African populations, and may also have been the basis for development of fermented milk products in some of these areas.

Even individual family size is influenced to a large extent by social or economic pressures. Large families were an asset in early years of farming and land development. In an industrialized and predominantly urban society, family planning is exerting a significant social force with genetic implications. The genetic effects of family planning include a decline in age-dependent defects such as some chromosome aberrations that are related to maternal age and point mutations related to paternal age. Decline in Rh disease associated with birth order may be due in part to family planning.

There is no doubt that genetic diseases constitute a significant portion of the world health deficit, and the advantages to improvement of the human race are seldom questioned. The controversy exists between those who advocate improvement in the species by selective breeding and those who recommend providing a better environment. Improvement of the race through altering the genotype is termed *eugenics;* improvement of the human race by modifying the environment is called *euthenics*.

Eugenics

Galton, in the late 1800's, coined the term eugenics (which means wellborn) to describe the selective elimination of undesirable phenotypes. Proponents of this philosophy feel that man can and should have some conscious control over his evolution. Eugenics is essentially planned breeding designed to alter future generations. Selective breeding has been successfully used for many years by animal and plant breeders in developing superior food products. For many persons any discussion of controlling heredity creates visions of Hitler's interpretation and misuse of directed evolution, and for some racial groups it is a code word for genocide. Religious groups protest that it is tampering with God's creation. For others it evokes the picture of a science fiction novel where people are made or unmade to order.

Positive eugenics is the attempt to encourage reproduction among those individuals considered to possess superior or beneficial genotypes. Suggested means for accomplishing this purpose include selected assortive mating of individuals with what are considered to be superior traits. Other methods are the establishing of sperm banks with sperm from a small, selected number of donors to be frozen and used to impregnate a large number of suitable women, and cloning the cells of persons with outstanding traits to produce, asexually, replicas of these desirable individuals. Some of the qualifications deemed to be superior might be physical characteristics, socially desirable personality, superior intellect, as well as absence of genetically determined defects.

Negative eugenics is the discouragement or prohibition of reproduction among individuals who are considered to be physically or mentally handicapped. Voluntary or legal prohibition of reproduction by persons with these genotypes might be accomplished with marriage laws, sterilization, and abortion.

The arguments for and against the relative merits of eugenics will continue for years to come. The multiple problems inherent in either of these proposals are readily apparent. What is considered an undesirable characteristic? Who makes the decision regarding the merits of a specific trait: scientists? physicians? lawyers?

clergy? government? Should a diabetic genius be encouraged to transmit his genes because he is a genius or be prohibited because he is diabetic? Where is the line drawn for IQ level? Many a genius has been described as emotionally unstable or eccentric. Should an epileptic Tchaikowsky or a deaf Beethoven be encouraged or prevented from procreation? Will a trait that is considered undesirable today be beneficial at another time or place? What about a single-gene defect such as albinism or colorblindness?

Euthenics

An opposite point of view is taken by those who support euthenics, which advocates the modification of the environment to allow the genetically abnormal individual to live a relatively normal life. Examples of euthenic measures are prescription glasses for nearsighted persons and special schools for the deaf. Medical treatments such as special diets for children with some of the inborn errors of metabolism, hormone replacement such as insulin for diabetes and thyroid for cretinism, or special orthopedic appliances and prosthetic devices can be considered as environmental manipulation. Overt aggressive behavior such as that attributed to the XYY male might be controlled through direction to socially acceptable activities such as competitive sports as opposed to assault and battery.

The individual and society

To what extent is scientific achievement contributing to the problems of the individual and society? What will the rapid development of knowledge mean in terms of man's ability to determine his fate more rapidly than society is able to cope with it? There are many legal, moral, and ethical issues to be faced and questions to be answered in relation to man and his future. A recurring theme is that of the rights of the individual as opposed to the rights of society. When dealing with genetics it is not only the rights of the individual today that must be considered but the rights of future generations—individuals and societies. If the cost (biologic and economic) of individual rights must be assumed by the remainder of the population, it then ceases to be an individual matter. If society is paying the cost, society will demand the right to make the decisions.

Some questions and issues that may need answers in the near future are:

Should reproduction of genetically high-risk persons be prohibited?

How much freedom of choice should be allowed?

Should couples be allowed to choose the sex of each child?

Who is responsible for children born with genetic defects that are serious enough to require long medical or institutional care?

Where a hereditary defect is viewed as creating a threat to society, should legislation be employed to protect the interests of society even at the expense of individual freedom?

Should there be compulsory mating between certain individuals in order to benefit society?

Should there be prohibition of undesirable pairings or compulsory sterilization of those at risk to bear defective offspring?

Should the husband who consents to artificial insemination by a donor be legally required to support the child?

Is the wife who submits to artificial insemination by a donor legally guilty of adultery?

Will the donor of the sperm have any claim over the child?

What is the responsibility of the genetically defective individual to society?

Should screening procedures continue to be carried out since detection and treatment of individuals with genetic disease will tend to increase the frequency of that gene in the population?

If treatment of a disease is effective is the disease really deleterious?

Are mandatory screening tests justified?

Should hereditary screening be required even if some persons object to such procedures as an invasion of privacy?

Who has a right to know the results of a test for genetic disease?

Should differentiation be made between a disease that is incapacitating and one in which the affected person is self-sustaining?

Can an individual who knows he or she is a carrier of a defective gene or chromosome but refuses to share this information with other family members (especially wife or husband) be held liable if an affected child is born in the family?

Will a child be able to sue his parents because he inherited an undesirable trait?

Should society continue to support persons in mental institutions because of hereditary mental illness?

If on examination an individual is informed that he stands a greater risk of being a criminal and asks for psychiatric help should this help be provided at public expense?

Should the mentally ill criminal offender be acquitted on the grounds that his behavior is attributed to his genetic constitution and, therefore, he is not responsible?

Society will be faced with these and many more questions related to genetically determined characteristics and associated problems.

What is the outlook for society in the face of scientific advancements? Pessimists predict a race of genetic cripples, but optimists feel assured that natural selection will continue to operate with survival of those that best adapt to the changing environment—even an environment of man-made alterations including treatment for genetic disease.

REFERENCES

1. Carter, C. O.: Genetic counseling, Med. Clin. North Am. **53:**991, 1969.
2. Hirshhorn, D.: Practical and ethical problems in human genetics. Birth defects, original article series **8:**17, 1972.
3. Leonard, D. O., Chase, G. A., and Childs, B.: Genetic counseling; a con-sumer's view, New Eng. J. Med. **287:** 433, 1972.
4. McKusick, V. A.: Human genetics, ed. 3, Englewood Cliffs, N. J., 1969, Prentice-Hall, Inc., Chapter 10.
5. Motulsky, A. G.: Genetic therapy; a clinical geneticist's response. In Hamil-

ton, M., editor: The new genetics and the future of man, Grand Rapids, Mich., 1972, Wm. B. Eerdmans Publishing Co.

6. Motulsky, A. G., Fraser, G. B., and Felsenstein, J.: Public health and long-term genetic implications of intrauterine diagnosis and selective abortion. Birth defects, original article series **7:**22, 1971.

7. Porter, I. H.: Heredity and disease, 1968, The Blakiston Division, McGraw-Hill Book Co.

8. Roberts, J. A. F.: An introduction to medical genetics, ed. 5, London, 1970, Oxford University Press, Inc., Chapter 12.

9. Thompson, J. S., and Thompson, M. W.: Genetics in medicine, Philadelphia, 1966, W. B. Saunders Co.

10. Watson, W., and Cann, H.: Genetic counseling in dermatology, Pediatr. Clin. North Am. **18:**757, 1971.

GENERAL REFERENCES

Carter, C. O.: An ABC of medical genetics, Boston, 1969, Little, Brown and Co., Chapter 7.

Fraser, F. C.: Taking the family history, Am. J. Med. **34:**585, 1963.

Fuhrmann, W., and Vogel, F.: Genetic counseling, New York, 1969, Springer-Verlag New York Inc.

Gordon, H.: Genetic counseling, JAMA **9:** 1215, 1971.

Hamilton, M., editor: The new genetics and the future of man, Grand Rapids, Mich., 1972, Wm. B. Eerdmans Publishing Co.

Hecht, F., and Lovrien, E. W.: Genetic diagnosis in the newborn, Pediatr. Clin. North Am. **17:**1039, 1970.

Hershey, N.: Legal and social policy issues pertaining to recent developments in genetics. Birth defects, original article series **8:**17, 1972.

Howell, R. R.: Prenatal diagnosis in the prevention of handicapping disorders, Pediatr. Clin. North Am. **20:**141, 1973.

Hsia, D. Y-Y.: The detection of heterozygous carriers, Med. Clin. North Am. **53:**857, 1969.

Nadler, H. L.: Indications of amniocentesis in the early prenatal detection of genetic disorders. Birth defects, original article series **7:**5, 1971.

Neel, J. V.: Thoughts on the future of human genetics, Med. Clin. North Am. **53:**1001, 1969.

Reed, S. C.: Counseling in medical genetics, ed. 2, Philadelphia, 1965, W. B. Saunders Co.

Reisman, L. E., and Matheny, A. P., Jr.: Genetics and counseling in medical practice, St. Louis, 1969, The C. V. Mosby Co.

Schulman, J. D.: Present status and future trends in prenatal diagnosis of metabolic disorders, Clin. Obstet. Gynecol. **15:**249, 1972.

Scriver, C. R.: Treatment of inherited disease: realized and potential, Med. Clin. North Am. **53:**941, 1969.

Shapiro, L. R.: The cytogenetics laboratory in pediatrics, Pediatr. Clin. North Am. **18:**209, 1971.

Stevenson, A. C., Davison, B. C. C., and Oakes, M. W.: Genetic counseling, Philadelphia, 1970, J. B. Lippincott Co.

Sultz, H. A., Schlesinger, E. R., and Feldman, J.: An epidemiologic justification for genetic counseling in family planning, Am. J. Pub. Health **62:**1489, 1972.

Glossary

Allele (allelomorph)—the different, or alternate, forms of a gene. Alleles that occur at the same position, or locus, on a chromosome pair may produce different effects during development.

Aneuploid—a chromosome number that deviates from the normal for a species. It may be more or less than the diploid number but is not an exact multiple of the basic haploid number found in germ cells.

Antibody—serum protein formed by immunocompetent cells in response to antigenic stimulation that reacts specifically with that antigen.

Antigen—a substance that is capable of stimulating an immune response.

Assortment—the random distribution of nonhomologous chromosomes to the germ cells during meiosis.

Autoimmunity—the production of antibodies against one's own tissues; immunity to self.

Autosome—chromosomes other than the sex (X or Y) chromosomes. Humans carry 22 pairs of autosomes.

Barr body or sex chromatin—a dark-stained mass located against the inner surface of the nuclear membrane in cells containing two or more X chromosomes. It is thought to represent the inactive X chromosome. Nuclei of normal female cells contain one Barr body; there is no Barr body in the nuclei of the normal male.

Carrier—an individual who possesses and can transmit the gene for a given trait but does not exhibit the trait himself.

Centromere—the small constricted region on a chromosome at which identical chromatids are joined and by which the chromosome is attached to a spindle fiber. The position of the centromere determines the length of the arms of the chromosome and, for any particular chromosome, is constant in location.

Chromatids—the individual strands of a chromosome that result from lengthwise duplication of the chromosome. When the centromere that joins the two chromatids separates, each chromatid becomes a separate chromosome.

Chromosome aberration—alteration from normal structure or number.

Chromosomes—various-sized structural elements in the cell nucleus, composed of deoxyribonucleic acid (DNA) and proteins, which carry the genes that convey the genetic information. Chromosomes have a species specific morphology and number.

Clone—a line of cells derived by asexual reproduction from a single cell, thus containing the same genetic constitution.

Codominance—both alleles expressed in the heterozygote.

Concordant—both members of a pair having the same trait or traits; used primarily in reference to twins.

Congenital—existing at birth as a result of heredity or prenatal environment.

Consanguinity—relationship of individuals having one or more ancestors in common.

Crossing over—exchange of comparable segments of genetic material between maternal and paternal homologous chromosomes during the first meiotic division.

Cytogenetics—a branch of genetics dealing with the study of chromosomes in relation to associated gene behavior.

Deletion—loss of a portion of a chromosome as a result of chromosome breakage.

Deoxyribonucleic acid (DNA)—the nucleic acid primarily contained within the cell nucleus, which carries the genetic information.

Dermatoglyphics—the patterns of skin ridges on fingers, toes, palms, and soles; the study of these patterns.

Differentiation—the process whereby embryonic cells acquire individual characteristics and function.

Diploid—possessing a paired set of chromosomes, one set from the father and one set from the mother (2n). Characteristic of all somatic cells.

Discordant—One member of a pair displaying a given trait while the other does not; a term usually used in twin studies.

Dizygotic—refers to twins derived from two fertilized eggs; fraternal twins.

Dominant—refers to a gene that produces an effect (is expressed) whenever it is present.

Drift (genetic)—the chance variation, extinction, or fixation of genes out of proportion to the frequency in the original population.

Duplication—a chromosome abnormality in which a segment of the chromosome is represented more than once as a result of breakage and refusion.

Embryo—the organism in the first stages of development. In humans this is generally considered to be the period from the end of the second week through the eighth week of gestation.

Eugenics—the science concerned with the improvement of the genetic potential of the human population by control of heredity through voluntary social action.

Euploid—a chromosome number that is the exact multiple of the haploid number.

Euthenics—the science concerned with improvement of the human race through control of environmental factors.

Expressivity—the degree to which an inherited trait is expressed in individuals with a particular genotype.

Fertilization—the union of a male gamete and a female gamete to form a zygote.

Fetus—the developing organism from the end of the eighth week of gestation until birth.

Gamete—a mature male or female germ cell (sperm or egg) containing a haploid number of chromosomes.

Gametogenesis—the orderly division of primary germ cells to produce gametes.

Gene—the basic unit of heredity. A finite segment of DNA that controls the production of a specific polypeptide.

Gene flow—a change of gene frequency in a population due to interbreeding or migration.

Genetic—dependent upon genes.

Genetic code—the triplet sequence of nucleotide bases in the DNA chain, as reflected in mRNA, that determines the sequence of amino acids during protein synthesis.

Genetic lethal—a gene or chromosome defect that renders the bearer unable to transmit the abnormality to the next generation.

Genetics—the science of inheritance and variation; the branch of biology concerned with the transmission of potential physical and mental characteristics from parents to offspring through the genes.

Genotype—the genetic constitution of an individual that determines the physical and chemical characteristics of that individual.

Germ cell—a gamete or the direct antecedent cell of a gamete.

Haploid—having a single set of unpaired chromosomes (N); characteristic of the gametes.

Hemizygous—having one unpaired gene or chromosome in an otherwise diploid cell; the X chromosome or genes carried on the X chromosome in the male.

Heredity, heritable—transmission of traits or characteristics from parent to offspring.

Heterogametic—producing two types of gametes. In humans the male produces an X-bearing and a Y-bearing gamete.

Heterozygous—having dissimilar genes at the same locus on homologous chromosomes.

Homogametic—producing only one type of gamete. In humans the female produces only X-bearing gametes.

Homologous chromosomes—structurally identical chromosomes containing the same types and number of genes.

Homozygous—having the same genes at a given locus on homologous chromosomes.

Immune reaction—the specific reaction between antigen and antibody.

Immunogenetics—the branch of genetics concerned with development of immunologic phenomena.

Inversion—a chromosomal abnormality in which a segment of the chromosome recombines in an inverted relationship following breakage.

Karyotype—a systematized array of chromosomes from a single cell, prepared by photography, that demonstrates the number and morphology of the chromosome complement.

Linkage—refers to genes located on the same chromosome.

Locus—the site occupied by a specific gene, or allele, on a particular chromosome.

Meiosis—the special type of cell division of gametogenesis by which a mother cell with a diploid set of chromosomes divides to produce gametes with a haploid, or single, set of chromosomes; reductional division.

Mitosis—somatic cell division by which the mother cell produces two daughter cells, each with the identical chromosome complement as the original cell.

Monosomy—a situation in which there is only one member of any given chromosome pair.

Monozygotic—refers to twins derived from a single fertilized ovum; identical twins.

Mosaicism—the presence in the same individual of two or more genotypically different cell lines.

Multifactorial—refers to inheritance determined by many factors with small additive effects.

Mutation—a hereditary change in genetic material. A mutation can be a change in a single gene (point mutation) or a change in chromosome characteristics.

Nondisjunction—Failure of two chromatids (during mitosis or meiosis II) or two homologous chromosomes (during meisis I) to separate or disjoin so that both pass to the same daughter cell.

Oogenesis—the series of cell divisions in the female germ cells that results in formation of functional ova; gametogenesis in the female.

Penetrance—the regularity with which a heritable trait is expressed in individuals carrying the gene for that trait.

Phenocopy—a condition or trait produced by environmental factors that is indistinguishable from one due to genetic factors.

Phenotype—the physical or chemical characteristics of an individual produced by interactions of the genotype with the environment.

Pleiotropy—the varied and multiple effects produced by a single gene.

Polygenic—inheritance involving many genes at separate loci whose combined, additive effects produce a given phenotype.

Polymorphism—the occurrence in a population of two or more genetically determined phenotypes in frequencies too great to be maintained by mutation.

Proband—an affected individual (irrespective of sex) through whom a family comes to the attention of an investigator; index case, propositus.

Propositus (female, proposita)—*see* proband.

Recessive—refers to a gene that produces its effect (is expressed) only when it is present in the homozygous state (double dose).

Ribonucleic acid (RNA)—the nucleic acid contained in the nucleus and the cytoplasm. In its various forms, RNA transfers genetic information within the cell.

Satellite—a small chromatin knob attached to the short arm of some chromosomes.

Second-set response—the rapid rejection of a tissue graft by a host already sensitized to the tissue of that particular genotype.

Segregation—the distribution of genes to gametes during the process of meiosis.

Selection—in population genetics, the operation of the forces that determine the biologic fitness of a genotype within a given population.

Sex chromatin—*see* Barr body.

Sex chromosomes—chromosomes that determine the sex of the individual. In humans, the X chromosomes in the female, the X and Y chromosomes in the male.

Sex influence—describes a trait carried on an autosome but expressed predominantly in one sex. When *only* one sex is affected the trait is said to be *sex limited*.

Sex-linked—refers to a gene located on a sex chromosome. Often used to describe genes on the X chromosome, although X-linked is the more accurate term.

Sex ratio—the ratio of males to females in a population.

Sibs (siblings)—brothers or sisters. A sibship is a group of brothers and sisters.

Spermatogenesis—the series of cell divisions in the male germ cells that result in formation of functional sperm; gametogenesis in the male.

Syndrome—a group of manifestations or symptoms that occur in association with each other.

Target tissue—in immunogenetics, a tissue against which antibodies are formed.

Translocation—the transfer of all or part of a chromosome to another location on the same chromosome or to a different chromosome following chromosome breakage.

Trisomy—a state in which there are three members of a given chromosome instead of the normal pair.

X chromosome—a sex chromosome found in duplicate in the normal female and singly in the male.

X-linkage—a gene carried on the X chromosome.

Y chromosome—one of the sex chromosomes found in the normal male. The Y chromosome is essential for the development of male gonads.

Zygote—the diploid cell formed by the fusion of the haploid male and female gametes; the fertilized ovum.

General references

Carter, C. O.: An ABC of medical genetics, Boston, 1969, Little, Brown and Co.

Clark, C. A.: Genetics for the clinician, ed. 2, Philadelphia, 1964, F. A. Davis Co.

Clark, C. A., editor: Selected topics in medical genetics, London, 1969, Oxford University Press, Inc.

Crispins, C. G., Jr.: Essentials of medical genetics, New York, 1971, Harper & Row, Publishers.

Emery, A. E. H.: Heredity, disease, and man, Berkeley, 1968, University of California Press.

Gardner, E. J.: Principles of genetics, ed. 4, New York, 1972, John Wiley & Sons, Inc.

Guyton, A. C.: Textbook of medical physiology, ed. 4, Philadelphia, 1971, W. B. Saunders Co.

Knudsen, A. G.: Genetics and disease, New York, 1965, The Blakiston Division, McGraw-Hill Book Co.

Levine, L.: Biology of the gene, ed. 2, St. Louis, 1973, The C. V. Mosby Co.

McKusick, V. A.: Human genetics, ed. 3, Englewood Cliffs, N. J., 1969, Prentice-Hall, Inc.

McKusick, V. A.: Mendelian inheritance in man, ed. 3, Baltimore, 1971, The Johns Hopkins University Press.

Moody, F. A.: Genetics of man, New York, 1967, W. W. Norton & Co., Inc.

Porter, I. H.: Heredity and disease, New York, 1968, The Blakiston Division, McGraw-Hill Book Co.

Reed, S. C.: Counseling in medical genetics, ed. 2, Philadelphia, 1965, W. B. Saunders Co.

Reisman, L. E., and Matheny, A. P., Jr.: Genetics and counseling in medical practice, St. Louis, 1969, The C. V. Mosby Co.

Roberts, J. A. F.: An introduction to medical genetics, ed. 5, London, 1970, Oxford University Press, Inc.

Stern, C.: Principles of human genetics, ed. 3, San Francisco, 1972, W. H. Freeman and Co. Publishers.

Stevenson, A. C., Davison, B. C. C., and Oakes, M. W.: Genetic counseling, Philadelphia, 1970, J. B. Lippincott Co.

Thompson, J. S., and Thompson, M. W.: Genetics in medicine, Philadelphia, 1966, W. B. Saunders Co.

Appendix A

The genetic code

The 64 triplet combinations of the genetic code with their amino acids

		Second letter				
		U	**C**	**A**	**G**	
First letter	**U**	UUU UUC } Phe UUA UUG } Leu	UCU UCC UCA UCG } Ser	UAU UAC } Tyr UAA } Ochre* UAG } Amber*	UGU UGC } Cys UGA* UGG } Tryp	U C A G
	C	CUU CUC CUA CUG } Leu	CCU CCC CCA CCG } Pro	CAU CAC } His CAA CAG } GluN	CGU CGC CGA CGG } Arg	U C A G
	A	AUU AUC } Ileu AUA } AUG Met†	ACU ACC ACA ACG } Thr	AAU AAC } AspN AAA AAG } Lys	AGU AGC } Ser AGA AGG } Arg	U C A G
	G	GUU GUC GUA GUG } Val	GCU GCC GCA GCG } Ala	GAU GAC } Asp GAA GAG } Glu	GGU GGC GGA GGG } Gly	U C A G

Third letter

*Ochre and amber are nonsense codons thought to be terminators. UGA is also a terminator.
†Methionine is an initiator.

U = uracil (RNA equivalent of thymine)   A = adenine
C = cytocine   G = guanine

205

The 20 Amino Acids

Alanine	Ala
Arginine	Arg
Asparagine	AspN
Aspartic acid	Asp
Cysteine	Cys
Glutamic acid	Glu
Glutamine	GluN
Glycine	Gly
Histidine	His
Isoleucine	Ileu
Leucine	Leu
Lysine	Lys
Methionine	Met
Phenylalanine	Phe
Proline	Pro
Serine	Ser
Threonine	Thr
Tryptophan	Tryp
Tyrosine	Tyr
Valine	Val

Appendix B

The blood group systems

The blood group systems

System	Year of discovery	Detectable antigens
ABO	1900	A,B,H*
MNSs	1927	M,N,S,s
P	1927	P_1,P_2,P^k
Rh	1940	C,C^w,c,D,d,E,e (Fisher and Race)
		$R^0,R^1,R^2,R^z,r,r',r'',r^y$ (Weiner)
Lutheran	1945	Lu^a,Lu^b
Kell	1946	K,k,Kp^a,Kp^b,Js^a,Js^b
Lewis	1946	Le^a,Le^b
Duffy	1950	Fy^a,Fy^b
Kidd	1951	Jk^a,Jk^b
Diego	1955	Di^a
I	1956	I,i
Auberger	1961	Au^a
Xg	1962	Xg^a
Dombrock	1965	Do^a
Stoltzfus	1969	Sf^a,Sf

*The H antigen is the precursor substance from which antigens A and B are formed by the action of A and B genes. It is detectable in type O, which contains unaltered H substance.

Appendix C

Dermatoglyphics

Every individual has an entirely different and distinct set of fingerprints. These complex, characteristic patterns, which are laid down during the third month of gestation and remain unchanged throughout a lifetime, have provided an invaluable means of personal identification. The patterns show a great deal of diversity in both detail and combinations of patterns. When the prints of any two individuals —even identical twins—are compared, numerous differences in configuration can be observed on fingers, palms, and soles. These patterns, or dermatoglyphics (*derm,* skin; *glyphein,* to carve), consist of epidermal ridges, creases, and cracks. The flexion creases are formed at the same time as the dermal ridges and, in some ways, affect the course of the ridges. Cracks, smaller than creases, tend to develop with age. Of significance to geneticists are the ridge prints and in some instances the creases.[1, 7]

Evidence that dermal ridge patterns are genetically determined is derived from similarities found among relatives, differences between unrelated individuals, and variations between members of different races. The diversity in types and combinations of patterns indicates that the dermal ridges are determined by many genes on many chromosomes. It has been logically assumed that the presence of extra genetic material that causes abnormal development in other organ systems might also produce alterations in dermal ridge patterns.

Palm lines (Fig. C-1).[5, 7] Preliminary dermatoglyphic analysis begins with natural centers, or *triradii.* A triradius is defined as the point at which three spokes, or main lines, meet and separate three regions, each of which contains almost parallel ridges. There are normally four triradii on the palm (digital triradii) at the base of each finger, labeled *a, b, c,* and *d;* and another, *t,* at the base of the palm near the wrist crease (axial triradius). The most prominent lines are designated by the capital letter corresponding to its triradius, for example, *A, D,* and so on. The position of the triradius can alter the *atd* angle; that is, the triangle formed by lines

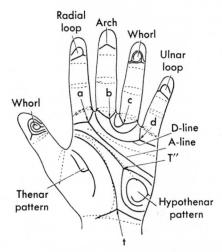

Fig. C-1. Dermatoglyphics on palms and fingertips, with nomenclature. (From Penrose, L. S.: Finger-prints, palms and chromosomes, Nature **197**:933, 1963.)

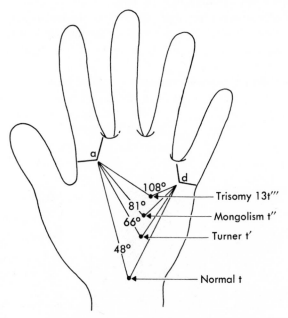

Fig. C-2. Mean position of most triradius *t* in children up to 4 years of age. (From Penrose, L. S.: Finger-prints, palms and chromosomes, Nature **197**:933, 1963.)

drawn from the *a* and *d* triradii at the base of the index and little fingers to the axial triradius (Fig. C-2). This angle is altered in chromosome abnormalities and also becomes narrower and longer with age.

Fingerprints. Fingerprint patterns are classified as arches, loops, and whorls (Fig. C-3). The classification is made on the basis of the number and position of triradii. For example, one triradius on a fingertip accompanies a loop pattern (loops may open toward the radial or ulnar side of the hand), two triradii are associated with a whorl; absence of a triradius produces an arch pattern. There are numerous variations on these basic patterns (double loops, tented arches, and such). One method of analyzing dermatoglyphic patterns is by a fingertip ridge count. The ridge counts are obtained by drawing a line from the triradius to the core of the print and counting the number of ridges it crosses (Fig. C-4). An average ridge count is 0 for arches, 12 for a loop, and 19 for a whorl. The total number of ridges for all ten fingers averages 145 in males and 127 in females.[5] In general, the more X chromosomes the fewer ridges. Also, Asiatics have more complex patterns (for example, whorls); Africans have simpler patterns.

Flexion creases. The palm normally shows three flexion creases (Fig. C-5, *A*) that are the heart, head, and life lines associated with palmistry. In some situations the two distal horizontal creases may be fused to form a single horizontal crease called a *single palmar crease,* or *simian crease* (Fig. C-5, *B*). Another variation noted by some investigators is the *Sydney line* in which the proximal transverse palmar crease extends to the ulnar margin of the palm (Fig. C-5, *C*).[3] Normally there are two flexion creases on each of the four fingers; however, occasionally the small finger contains only one.

Soles. The ridge patterns on the soles of the feet also show individualized patterns. The only one of clinical importance is the hallucal region, situated at the base of the great toe, which normally shows loops, whorls, and arches, both tibial and fibular.[6, 7]

Peculiarities in dermal patterns related to various clinical syndromes. Characteristic dermatoglyphic patterns have been noted in almost all of the chromosomal

Fig. C-3. Examples of the three basic fingerprint types. **A,** Arch; **B,** loop; **C,** whorl.

aberrations. The table on p. 212 indicates some of the patterns frequently associated with specific abnormalities. In addition, an increased incidence of characteristic palm creases have been observed in children with rubella syndrome and leukemia. For example, the Sydney line has been noted in a large percentage of children with rubella than in controls and a significant number (16%) of children with leukemia.[2, 4]

Fig. C-4. Hypothetical fingerprint pattern illustrating how a ridge count is made. The line on the left, joining the triradius to the core, crosses 12 ridges; the line on the right crosses 8 ridges.

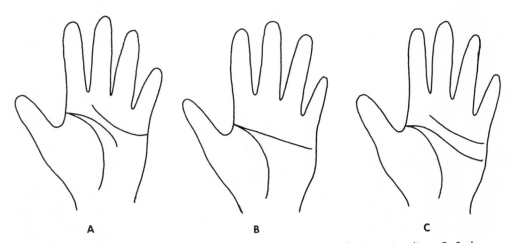

Fig. C-5. Examples of flexion creases on the palm. **A,** Normal; **B,** simian line; **C,** Sydney line.

Dermatoglyphic patterns and chromosome disorders[5, 7, 8]

Disorder	Fingers	Palm	Hallucal area of sole
Trisomy 21 G (Down's syndrome)	Excess ulnar loops (10 loops in 60%)	Single palmar crease	Arch tibial (50%) Small loop (35%)
Trisomy 18 E	Arch patterns	Distal axial triradius	Open field
Trisomy 13 D	Increased number of arches	Single palmar crease	Large pattern, loop tibial, and arch pattern
Cri du chat syndrome	Increased number of arches	Distal axial triradius	Open field
Turner's syndrome (XO)	Tendency toward high ridge counts, patterns usually large	Slightly distal axial triradius	Large pattern, loop distal
Klinefelter's syndrome (XXY)	Increased number of arches	More proximal axial triradius	Not characteristic
XXX female	Normal	Normal	Normal

REFERENCES

1. Alter, M.: Dermatoglyphic analysis as a diagnostic tool, Medicine **46:**35, 1967.
2. Alter, M.: Variation in palmar creases, Am. J. Dis. Child. **120:**424, 1970.
3. Johnson, C. F., and Opitz, E.: Unusual palm creases and unusual children, Clin. Pediatr. **12:**101, 1973.
4. Menser, M. A., and Purvis-Smith, S. G.: Dermatoglyphic defects in children with leukemia, Lancet **1:**1076, 1969.
5. Penrose, L. S.: Finger-prints, palms and chromosomes, Nature **197:**933, 1963.
6. Penrose, L. S.: Dermatoglyphic topology, Nature **205:**544, 1965.
7. Thompson, J. S., and Thompson, M. W.: Genetics in medicine, Philadelphia, 1966, W. B. Saunders Co., Chapter 14.
8. Uchida, I. A., and Soltan, H. D.: Evaluation of dermatoglyphics in medical genetics, Pediatr. Clin. North Am. **10:**409, 1963.

Index